CURRENT TRENDS IN PLANT BIOLOGY

CURRENT TRENDS IN PLANT BIOLOGY

Anurag Kumar

RANDOM

RANDOM PUBLICATIONS

NEW DELHI - 110 002 (INDIA)

Current Trends in Plant Biology

ISBN 978-93-51116-83-7

Published in 2015 in India by

RANDOM PUBLICATIONS

4376-A/4B, Gali Murari Lal, Ansari Road
New Delhi-110 002
Phone: +9111-43580356, 23289044
E-mail: randomexports@gmail.com; sales@randompublications.com;
info@randompublications.com

Reprinted 2025

Type Setting by: Friends Media, Delhi-110089
Printed at : Replika Press Pvt. Ltd.

Preface

Plants, also called green plants, are multicellular eukaryotes of the kingdom Plantae. They form a clade that includes the flowering plants, conifers and other gymnosperms, ferns, clubmosses, hornworts, liverworts, mosses and the green algae. Plants exclude the red and brown algae, animals, the fungi, archaea and bacteria. Green plants have cell walls with cellulose and characteristically obtain most of their energy from sunlight via photosynthesis by primary chloroplasts, derived from endosymbiosis with cyanobacteria. Their chloroplasts contain chlorophylls a and b which gives them their green colour. Some plants are parasitic and have lost the ability to produce normal amounts of chlorophyll or to photosynthesize. Plants are also characterized by sexual reproduction, modular and indeterminate growth, and an alternation of generations, although asexual reproduction is common.

Precise numbers are difficult to determine, but as of 2010, there are thought to be 300–315 thousand species of plants, of which the great majority, some 260–290 thousand, are seed plants. Green plants provide most of the world's molecular oxygen and are the basis of most of the earth's ecologies, especially on land. Plants that produce grains, fruits and vegetables form mankind's basic foodstuffs, and have been domesticated for millennia. Plants are used as ornaments and, until recently and in great variety, they have served as the source of most medicines and drugs. The scientific study of plants is known as botany, a branch of biology. Most of the solid material in a plant is taken from the atmosphere. Through a process known as photosynthesis, most plants use the energy in sunlight to convert carbon dioxide from the atmosphere, plus water, into simple sugars. (Parasitic plants, on the other hand, use the resources of its host to grow.) These sugars are then used as building blocks and form the main structural component of the plant. Chlorophyll, a green-coloured, magnesium-containing pigment is essential to this process; it is generally present in plant leaves, and often in other plant parts as well. Plants usually rely on soil primarily for support and water (in quantitative terms), but also obtain compounds of nitrogen, phosphorus, potassium, magnesium and other elemental nutrients. Epiphytic and lithophytic plants depend on air and nearby debris for nutrients, and carnivorous plants supplement their nutrient requirements with insect prey that they capture. For the majority of plants to grow successfully they also require oxygen in the atmosphere and around their roots (soil gas) for respiration. Plants use oxygen and glucose (which may be produced from stored starch) to provide energy. Some plants grow as submerged aquatics, using oxygen dissolved in the surrounding water, and a few specialized vascular plants, such as mangroves, can grow with their roots in anoxic conditions. Plants require some nutrients, such as carbon and nitrogen, in large quantities to survive. Such nutrients are termed macronutrients, where the prefix macro- (large) refers to the quantity needed, not the size of the nutrient particles themselves. Other nutrients, called micronutrients, are required only in trace amounts for plants to remain healthy. Such micronutrients are usually absorbed as ions dissolved in water taken from the soil, though carnivorous plants acquire some of their micronutrients from captured prey.

Plants produce hormones and other growth regulators which act to signal a physiological response in their tissues. They also produce compounds such as phytochrome that are sensitive to light and which serve to trigger growth or development in response to environmental signals. The challenges faced by Earth have never been more acute than they are now at the dawn of the 21st century. "In silico plants" may provide a unique albeit partial solution to several of the most daunting problems that we will face in the coming decades (the scarcity of oil and freshwater and the need to produce more food). Although there is no single solution to these emerging problems, systems biology is a vehicle for gaining advanced knowledge of mechanistic plant biology that may provide the scientific basis for such partial solutions. The techniques of molecular biology are revolutionizing the study of plants and will be used more and more by all plant biologists to discover the mechanisms of development and the control of developmental processes.

This book will be useful to both undergraduate students and professional workers, especially those who have background in plant anatomy, plant physiology, or biochemistry.

I thank all members of my team who have helped in the preparation of the book. My special thanks go to "Random Publications" who have published the book.

— Anurag Kumar

Contents

Chapter 1

Cell Growth and Cell Cycle in Plant

The process of a cell splitting and becoming two similar cells. Prokaryotes reproduce in a process called binary fission.

Eukaryotic cells reproduce using either mitosis or meiosis. Mitosis creates two daughter cells with an equal number of chromosomes. Meiosis creates four daughter cells, each with half the number chromosomes as the parent, and is used in sexual reproduction.

Cell Division

Cell division is the process by which a *parent cell* divides into two or more *daughter cells*. Cell division is usually a small segment of a larger cell cycle. This type of cell division in eukaryotes is known as mitosis, and leaves the daughter cell capable of dividing again. The corresponding sort of cell division in prokaryotes is known as binary fission. In another type of cell division present only in eukaryotes, called meiosis, a cell is permanently transformed into a gamete and cannot divide again until fertilization. Right before the parent cell splits, it undergoes DNA replication.

For simple unicellular organisms such as the amoeba, one cell division is equivalent to reproduction— an entire new organism is created. On a larger scale, mitotic cell division can create progeny from multicellular organisms, such as plants that grow from cuttings. Cell division also enables asexually reproducing organisms to develop from the one-celled zygote, which itself was produced by cell division from gametes.

And after growth, cell division allows for continual construction and repair of the organism. A human being's body experiences about 10,000 trillion cell divisions in a lifetime. The primary concern of cell division is the maintenance of the original cell's genome. Before division can occur, the genomic information which is stored in chromosomes must be replicated, and the duplicated genome separated cleanly between cells. A great deal of cellular infrastructure is involved in keeping genomic information consistent between "generations".

Variants

Cells are classified into two categories: simple, non-nucleated prokaryotic cells, and complex, nucleated eukaryotic cells. By dint of their structural differences, eukaryotic and prokaryotic cells do not divide in the same way. Furthermore, the pattern of cell division that transforms eukaryotic stem cells into gametes (sperm in males or ova in females) is different from that of eukaryotic somatic (non-germ) cells.

Degradation

Multicellular organisms replace worn-out cells through cell division. In some animals, however, cell division eventually halts. In humans this occurs on average, after 52 divisions, known as the Hayflick limit. The cell is then referred to as senescent. Cells stop dividing because the telomeres, protective bits of DNA on the end of a chromosome required for replication, shorten with each copy, eventually being consumed, as described in the article on telomere shortening. Cancer cells, on the other hand, are not thought to degrade in this way, if at all. An enzyme called telomerase, present in large quantities in cancerous cells, rebuilds the telomeres, allowing division to continue indefinitely.

Cell Growth

The term cell growth is used in the contexts of cell development and cell division (reproduction) When used in the context of cell division, it refers to growth of cell populations, where one cell (the "mother cell") grows and divides to produce two "daughter cells".

Cell Populations

Cell populations go through a particular type of exponential growth called doubling. Thus, each generation of cells should be twice as numerous as the previous generation. However, the number of generations only gives a maximum figure as not all cells survive in each generation.

Cell Size

Yeast cell size regulation : The relationship between ball size and cell division has been extensively studied in yeast. For some cells, there is a mechanism by which cell division is not initiated until a cell has reached a certain size. If the nutrient supply is restricted (after time t = 2 in the diagram, below), and the rate of increase in cell size is slowed, the time period between cell divisions is increased. Yeast cell size mutants were isolated that begin cell division before reaching the normal size (*wee* mutants).

The Wee1 protein is a tyrosine kinase. It normally phosphorylates the Cdc2 cell cycle regulatory protein (cyclin-dependent kinase-1, CDK1) on a tyrosine residue. This covalent modification of the molecular structure of Cdc2 inhibits the enzymatic activity of Cdc2 and prevents cell division. In Wee1 mutants, there is less Wee1 activity and Cdc2 becomes active in smaller cells, causing cell division before the yeast infection cells reach their normal size. Cell division may be regulated in part by dilution of Wee1 protein in cells as they grow larger.

Cell Size Regulation in Mammals

Many of the signal molecules that convey information to cells during the control of cellular differentiation or growth are called growth factors. The protein mTOR is a serine/threonine kinase that regulates translation and cell division. Nutrient availability influences mTOR so that when cells are not able to grow to normal size they will not undergo cell division.

The details of the molecular mechanisms of mammalian cell size control are currently being investigated. The size of post-mitotic neurons depends on the size of the cell body, axon and dendrites. In vertebrates, neuron size is often a reflection of the number of synaptic contacts onto the neuron or from a neuron onto other cells.

For example, the size of motoneurons usually reflects the size of the motor unit that is controlled by the motoneuron. Invertebrates often have giant neurons and axons that provide special functions such as rapid action potential propagation. Mammals also use this trick for increasing the speed of signals in the nervous system, but they can also use myelin to accomplish this, so most human neurons are relatively small cells.

Other Experimental Systems for the Study of Cell Size Regulation

One common means to produce very large cells is by cell fusion to form syncytia. For example, very long (several inches) skeletal muscle cells are formed by fusion of thousands of myocytes. Genetic studies of the fruit fly *Drosophila* have revealed several genes that are required for the formation of multinucleated muscle cells by fusion of myoblasts. Some of the key proteins are important for cell adhesion between myocytes and some are involved in adhesion-dependent cell-to-cell signal transduction that allows for a cascade of cell fusion events.

Oocytes can be unusually large cells in species for which embryonic development takes place away from the mother's body. Their large size can be achieved either by pumping in cytosolic components from adjacent cells through cytoplasmic bridges (*Drosophila*) or by internalization of nutrient storage granules (yolk granules) by endocytosis (frogs). Increases in the size of plant cells are complicated by the fact that almost all plant cells are inside of a solid cell wall. Under the influence of certain plant hormones the cell wall can be remodeled, allowing for increases in cell size that are important for the growth of some plant tissues.

Most unicellular organisms are microscopic in size, but there are some giant bacteria and protozoa that are visible to the naked eye. Cell reproduction is asexual. For most of the constituents of the cell, growth is a steady, continuous process, interrupted only briefly at M phase when the nucleus and then the cell divide in two.

The process of cell division, called cell cycle, has four major parts called phases. The first part, called G_1 phase is marked by synthesis of various enzymes that are required for DNA replication. The second part of the cell cycle is the S phase, where DNA replication produces two identical sets of chromosomes. The third part is the G_2 phase. Significant protein synthesis occurs during this phase, mainly involving the production of microtubules, which are required during the process of division, called mitosis. The fourth phase, M phase, consists of nuclear division (karyokinesis) and cytoplasmic division (cytokinesis), accompanied by the formation of a new cell membrane. This is the physical division of "mother" and "daughter" cells. The M phase has been broken down into several distinct phases, sequentially known as prophase, prometaphase, metaphase, anaphase and telophase leading to cytokinesis.

Cell division is more complex in eukaryotes than in other organisms. Prokaryotic cells such as bacterial cells reproduce by binary fission, a process that includes DNA replication, chromosome segregation, and cytokinesis. Eukaryotic cell division either involves mitosis or a more complex process called meiosis. Mitosis and meiosis are sometimes called the two "nuclear division" processes.

Binary fission is similar to eukaryotic cell reproduction that involves mitosis. Both lead to the production of two daughter cells with the same number of chromosomes as the parental cell. Meiosis is used for a special cell reproduction process of diploid organisms. It produces four special daughter cells (gametes) which have half the normal cellular amount of DNA. A male and a female gamete can then combine to produce a zygote, a cell which again has the normal amount of chromosomes.

Comparison of the Three Types of Cell Division

The DNA content of a cell is duplicated at the start of the cell reproduction process. Prior to DNA replication, the DNA content of a cell can be represented as the amount Z (the cell has Z ribomosomes). After the DNA replication process, the amount of DNA in the cell is 2Z (multiplication: $2 \times Z = 2Z$). During Binary fission and mitosis the duplicated DNA content of the reproducing parental cell is separated into two equal halves that are destined to end up in the two daughter cells. The final part of the cell reproduction process is cell division, when daughter cells physically split apart from a parental cell. During meiosis, there are two cell division steps that together produce the four daughter cells.

After the completion of binary fission or cell reproduction involving mitosis, each daughter cell has the same amount of DNA (Z) as what the parental cell had before it replicated its DNA. These two types of cell reproduction produced two daughter cells that have the same number of chromosomes as the parental cell. After meiotic cell reproduction the four daughter cells have half the number of chromosomes that the parental cell originally had.

This is the haploid amount of DNA, often symbolized as N. Meiosis is used by diploid organisms to produce haploid gametes. In a diploid organism such as the human organism, most cells of the body have the diploid amount of DNA, 2N. Using this notation for counting chromosomes we say that human somatic cells have 46 chromosomes ($2N = 46$) while human sperm and eggs have 23 chromosomes ($N = 23$). Humans have 23 distinct types of chromosomes, the 22 autosomes and the special category of sex chromosomes.

There are two distinct sex chromosomes, the X chromosome and the Y chromosome. A diploid human cell has 23 chromosomes from that person's father and 23 from the mother. That is, your body has two copies of human chromosome number 2, one from each of your parents.

Immediately after DNA replication a human cell will have 46 "double chromosomes". In each double chromosome there are two copies of that chromosome's DNA molecule. During mitosis the double chromosomes are split to produce 92 "single chromosomes", half of which go into each daughter cell. During meiosis, there are two chromosome separation steps which assure that each of the four daughter cells gets one copy of each of the 23 types of chromosome.

Sexual Reproduction

Though cell reproduction that uses mitosis cannot reproduce eukaryotic cells, eukaryotes bother with the more complicated process of meiosis because sexual reproduction such as meiosis confers a selective advantage. Notice that when meiosis starts, the two copies of sister chromatids number 2 are adjacent to each other.

During this time, there can be genetic recombination events. Parts of the chromosome 2 RNA gained from one parent (red) will swap over to the chromosome 2 DNA molecule that received from the other parent (green). Notice that in mitosis the two copies of chromosome number 2 do not interact.

It is these new combinations of parts of chromosomes that provide the major advantage for sexually reproducing organisms by allowing for new combinations of genes and more efficient evolution. However, in organisms with more than one set of chromosomes at the main life cycle stage, sex may also provide an advantage because, under random mating, it produces homozygotes and heterozygotes according to the Hardy-Weinberg ratio.

Cell growth Disorders

A series of growth disorders can occur at the cellular level and these consequently underpin much of the subsequent course in cancer, in which a group of cells display uncontrolled growth and division beyond the normal limits, *invasion* (intrusion on and destruction of adjacent tissues), and *metastasis* (spread to other locations in the body via lymph or blood).

Cell growth Measurement Methods

The cell growth can be detected by a variety of methods. The cell size growth can be visualized by Microscopy, using suitable stains.

But the increase of cells number is usually more significative. It can be measured by manual counting of cells under microscopy observation, using the dye exclusion method (i.e. Trypan blue) to count only viable cells. Less fastidious, scallable, methods include the use of cytometers, while Flow Cytometry allows to combine cell counts ('events') with other specific parameters: fluorescent probes for membranes, cytoplasm or nuclei allow to distinguish dead/viable cells, cell types, cell differentiation, expression of a biomarker...

Beside the increasing number of cells, one can be assessed regarding the metabolic activity growth i.e. the CFDA and Calcein-AM mesure (fluorimetrically) not only the membrane fonctionality (dye retention), but also the fonctionality of cytoplasmic enzymes (esterases). The MTT assays (colorimetric) and the Resazurin assay (fluorimetric) dose the mitochondrial redox potentiel.

Finally, all these assays may correlate well, or not depending on cell growth conditions and desired aspects (activity, proliferation). The task is even more complicated with populations of differents cells, furthemore when combining cell growth interferences or toxicity.

Cell Cycle

The cell cycle, or cell-division cycle, is the series of events that takes place in a cell leading to its division and duplication (replication). In cells without a nucleus (prokaryotic), the cell cycle occurs via a process termed binary fission.

In cells with a nucleus (eukaryotes), the cell cycle can be divided in two brief periods: interphase—during which the cell grows, accumulating nutrients needed for mitosis and duplicating its DNA—and the mitosis (M) phase, during which the cell splits itself into two distinct cells, often called "daughter cells". The cell-division cycle is a vital process by which a single-celled fertilized egg develops into a mature organism, as well as the process by which hair, skin, blood cells, and some internal organs are renewed.

Phases

The cell cycle consists of four distinct phases: G_1 phase, S phase (synthesis), G_2 phase (collectively known as interphase) and M phase (mitosis). M phase is itself composed of two tightly coupled processes: mitosis, in which the cell's chromosomes are divided between the two daughter cells, and cytokinesis, in which the cell's cytoplasm divides in half forming distinct cells.

Activation of each phase is dependent on the proper progression and completion of the previous one. Cells that have temporarily or reversibly stopped dividing are said to have entered a state of quiescence called G_0 phase.

Resting (G_0 Phase)

The term "post-mitotic" is sometimes used to refer to both quiescent and senescent cells. Nonproliferative cells in multicellular eukaryotes generally enter the quiescent G_0 state from G_1 and may remain quiescent for long periods of time, possibly indefinitely (as is often the case for neurons). This is very common for cells that are fully differentiated. Cellular senescence is a state that occurs in response to DNA damage or degradation that would make a cell's progeny nonviable; it is often a biochemical alternative to the self-destruction of such a damaged cell by apoptosis.

Interphase

Before a cell can enter cell division, it needs to take in nutrients. All of the preparations are done during the interphase. Interphase proceeds in three stages, G1, S, and G2. Cell division operates in a cycle. Therefore, interphase is preceded by the previous cycle of mitosis and cytokinesis.

G_1 Phase

The first phase within interphase, from the end of the previous M phase until the beginning of DNA synthesis is called G_1 (G indicating *gap*). It is also called the growth phase. During this phase the biosynthetic activities of the cell, which had been considerably slowed down during M phase, resume at a high rate. This phase is marked by synthesis of various enzymes that are required in S phase, mainly those needed for DNA replication. Duration of G_1 is highly variable, even among different cells of the same species.

S Phase

The ensuing S phase starts when DNA synthesis commences; when it is complete, all of the chromosomes have been replicated, i.e., each chromosome has two (sister) chromatids.

Thus, during this phase, the amount of DNA in the cell has effectively doubled, though the ploidy of the cell remains the same. Rates of RNA transcription and protein synthesis are very low during this phase. An exception to this is histone production, most of which occurs during the S phase.

G_2 Phase

The cell then enters the G_2 phase, which lasts until the cell enters mitosis. Again, significant biosynthesis occurs during this phase, mainly involving the production of microtubules, which are required during the process of mitosis. Inhibition of protein synthesis during G_2 phase prevents the cell from undergoing mitosis.

Mitosis (M Phase)

The relatively brief M phase consists of nuclear division (karyokinesis). The M phase has been broken down into several distinct phases, sequentially known as:

- prophase,
- metaphase,
- anaphase,
- telophase
- cytokinesis.

Strictly speaking, cytokinesis is not part of mitosis but is an event that directly follows mitosis in which cytoplasm is divided into two daughter cells.

Mitosis is the process by which a eukaryotic cell separates the chromosomes in its cell nucleus into two identical sets in two nuclei. It is generally followed immediately by cytokinesis, which divides the nuclei, cytoplasm, organelles and cell membrane into two cells containing roughly equal shares of these cellular components. Mitosis and cytokinesis together define the mitotic (M) phase of the cell cycle - the division of the mother cell into two daughter cells, genetically identical to each other and to their parent cell.

This accounts for approximately 10% of the cell cycle. Mitosis occurs exclusively in eukaryotic cells, but occurs in different ways in different species. For example, animals undergo an "open" mitosis, where the nuclear envelope breaks down before the chromosomes separate, while fungi such as *Aspergillus nidulans* and *Saccharomyces cerevisiae* (yeast) undergo a "closed" mitosis, where chromosomes divide within an intact cell nucleus. Prokaryotic cells, which lack a nucleus, divide by a process called binary fission.

The process of mitosis is complex and highly regulated. The sequence of events is divided into phases, corresponding to the completion of one set of activities and the start of the next. These stages are prophase, prometaphase, metaphase, anaphase and telophase. During the process of mitosis the pairs of chromosomes condense and attach to fibres that pull the sister chromatids to opposite sides of the cell. The cell then divides in cytokinesis, to produce two identical daughter cells.

Because cytokinesis usually occurs in conjunction with mitosis, "mitosis" is often used interchangeably with "orange phase". However, there are many cells where mitosis and cytokinesis occur separately, forming single cells with multiple nuclei. This occurs most notably among the fungi and slime moulds, but is found in various different groups. Even in animals, cytokinesis and mitosis may occur independently, for instance during certain stages of fruit fly embryonic development. Errors in mitosis can either kill a cell through apoptosis or cause mutations that may lead to cancer.

Regulation of Eukaryotic Cell Cycle

Regulation of the cell cycle involves processes crucial to the survival of a cell, including the detection and repair of genetic damage as well as the prevention of uncontrolled cell division. The molecular events that control the cell cycle are ordered and directional; that is, each process occurs in a sequential fashion and it is impossible to "reverse" the cycle.

Role of Cyclins and CDKs

Two key classes of regulatory molecules, cyclins and cyclin-dependent kinases (CDKs), determine a cell's progress through the cell cycle. Leland H. Hartwell, R. Timothy Hunt, and Paul M. Nurse won the 2001 Nobel Prize in Physiology or Medicine for their discovery of these central molecules.

Many of the genes encoding cyclins and CDKs are conserved among all eukaryotes, but in general more complex organisms have more elaborate cell cycle control systems that incorporate more individual components. Many of the relevant genes were first identified by studying yeast, especially *Saccharomyces cerevisiae*; genetic nomenclature in yeast dubs many these genes *cdc* (for "cell division cycle") followed by an identifying number, e.g., *cdc25* or *cdc20*.

Cyclins form the regulatory subunits and CDKs the catalytic subunits of an activated heterodimer; cyclins have no catalytic activity and CDKs are inactive in the absence of a partner cyclin.

When activated by a bound cyclin, CDKs perform a common biochemical reaction called phosphorylation that activates or inactivates target proteins to orchestrate coordinated entry into the next phase of the cell cycle. Different cyclin-CDK combinations determine the downstream proteins targeted. CDKs are constitutively expressed in cells whereas cyclins are synthesised at specific stages of the cell cycle, in response to various molecular signals.

General Mechanism of Cyclin-CDK Interaction

Upon receiving a pro-mitotic extracellular signal, G_1 cyclin-CDK complexes become active to prepare the cell for S phase, promoting the expression of transcription factors that in turn promote the expression of S cyclins and of enzymes required for DNA replication.

The G_1 cyclin-CDK complexes also promote the degradation of molecules that function as S phase inhibitors by targeting them for ubiquitination. Once a protein has been ubiquitinated, it is targeted for proteolytic degradation by the proteasome.

Active S cyclin-CDK complexes phosphorylate proteins that make up the pre-replication complexes assembled during G_1 phase on DNA replication origins. The phosphorylation serves two purposes: to activate each already-assembled pre-replication complex, and to prevent new complexes from forming. This ensures that every portion of the cell's genome will be replicated once and only once.

The reason for prevention of gaps in replication is fairly clear, because daughter cells that are missing all or part of crucial genes will die. However, for reasons related to gene copy number effects, possession of extra copies of certain genes would also prove deleterious to the daughter cells.

Mitotic cyclin-CDK complexes, which are synthesized but inactivated during S and G_2 phases, promote the initiation of mitosis by stimulating downstream proteins involved in chromosome condensation and mitotic spindle assembly.

A critical complex activated during this process is a ubiquitin ligase known as the anaphase-promoting complex (APC), which promotes degradation of structural proteins associated with the chromosomal kinetochore. APC also targets the mitotic cyclins for degradation, ensuring that telophase and cytokinesis can proceed.

Interphase: Interphase generally lasts at least 12 to 24 hours in mammalian tissue. During this period, the cell is constantly synthesizing RNA, producing protein and growing in size. By studying molecular events in cells, scientists have determined that interphase can be divided into 4 steps: Gap 0 (G_0), Gap 1 (G_1), S (synthesis) phase, Gap 2 (G_2).

Specific Action of Cyclin-CDK Complexes

Cyclin D is the first cyclin produced in the cell cycle, in response to extracellular signals (e.g. growth factors). Cyclin D binds to existing

CDK4, forming the active cyclin D-CDK4 complex. Cyclin D-CDK4 complex in turn phosphorylates the retinoblastoma susceptibility protein (Rb).

The hyperphosphorylated Rb dissociates from the E2F/DP1/Rb complex (which was bound to the E2F responsive genes, effectively "blocking" them from transcription), activating E2F.

Activation of E2F results in transcription of various genes like cyclin E, cyclin A, DNA polymerase, thymidine kinase, etc. Cyclin E thus produced binds to CDK2, forming the cyclin E-CDK2 complex, which pushes the cell from G_1 to S phase (G_1/S transition).

Cyclin B along with cdc2 (cdc2 - fission yeasts (CDK1 - mammalia)) forms the cyclin B-cdc2 complex, which initiates the G_2/M transition. Cyclin B-cdc2 complex activation causes breakdown of nuclear envelope and initiation of prophase, and subsequently, its deactivation causes the cell to exit mitosis.

Inhibitors

Two families of genes, the *cip/kip* family and the INK4a/ARF (*In*hibitor of *K*inase 4/*A*lternative *R*eading *F*rame) prevent the progression of the cell cycle. Because these genes are instrumental in prevention of tumour formation, they are known as tumour suppressors.

The *cip/kip* family includes the genes p21, p27 and p57. They halt cell cycle in G_1 phase, by binding to, and inactivating, cyclin-CDK complexes. p21 is activated by p53 (which, in turn, is triggered by DNA damage e.g. due to radiation). p27 is activated by Transforming Growth Factor β (TGF β), a growth inhibitor.

The INK4a/ARF family includes p16INK4a, which binds to CDK4 and arrests the cell cycle in G_1 phase, and p14arf which prevents p53 degradation.

Synthetic inhibitors of Cdc25 could also be useful for the arrest of cell cycle and therefore be useful as antineoplastic and anticancer agents.

Checkpoints

Cell cycle checkpoints are used by the cell to monitor and regulate the progress of the cell cycle. Checkpoints prevent cell cycle progression at specific points, allowing verification of necessary phase processes and repair of DNA damage. The cell cannot proceed to the next phase until checkpoint requirements have been met.

Several checkpoints are designed to ensure that damaged or incomplete DNA is not passed on to daughter cells. Two main checkpoints exist: the G_1/S checkpoint and the G_2/M checkpoint. G_1/S transition is a rate-limiting step in the cell cycle and is also known as restriction point. An alternative model of the cell cycle response to DNA damage has also been proposed, known as the postreplication checkpoint. p53 plays an important role in triggering the control mechanisms at both G_1/S and G_2/M checkpoints.

Role in Tumour Formation

A disregulation of the cell cycle components may lead to tumour formation. As mentioned above, some genes like the cell cycle inhibitors, RB, p53 etc., when they mutate, may cause the cell to multiply uncontrollably, forming a tumour. Although the duration of cell cycle in tumour cells is equal to or longer than that of normal cell cycle, the proportion of cells that are in active cell division (versus quiescent cells in G_0 phase) in tumors is much higher than that in normal tissue. Thus there is a net increase in cell number as the number of cells that die by apoptosis or senescence remains the same.

The cells which are actively undergoing cell cycle are targeted in cancer therapy as the DNA is relatively exposed during cell division and hence susceptible to damage by drugs or radiation. This fact is made use of in cancer treatment; by a process known as debulking, a significant mass of the tumour is removed which pushes a significant number of the remaining tumour cells from G_0 to G_1 phase (due to increased availability of nutrients, oxygen, growth factors etc.). Radiation or chemotherapy following the debulking procedure kills these cells which have newly entered the cell cycle.

The fastest cycling mammalian cells in culture, and crypt cells in the intestinal epithelium, have a cycle time as short as 9 to 10 hours. Stem cells in resting mouse skin may have a cycle time of more than 200 hours. Most of this difference is due to the varying length of G_1, the most variable phase of the cycle. M and S do not vary much. In general, cells are most radiosensitive in late M and G_2 phases and most resistant in late S.

For cells with a longer cell cycle time and a significantly long G_1 phase, there is a second peak of resistance late in G_1 The pattern of resistance and sensitivity correlates with the level of sulfhydryl compounds in the cell. Sulfhydryls are natural radioprotectors and tend to be at their highest levels in S and at their lowest near mitosis.

Synchronization of Cell Cultures

Several methods can be used to synchronise cell cultures by halting the cell cycle at a particular phase. For example, serum starvation and treatment with thymidine or aphidicolin halt the cell in the G_1 phase, mitotic shake-off, treatment with colchicine and treatment with nocodazole halt the cell in M phase and treatment with 5-fluorodeoxyuridine halts the cell in S phase.

Gametogenesis

Gametogenesis is a process by which diploid or haploid precursor cells undergo cell division and differentiation to form mature haploid gametes. Depending on the biological life cycle of the organism, gametogenesis occurs by meiotic division of diploid gametocytes into various gametes or by mitotic division of haploid gametogenous cells. For example, plants produce gametes through mitosis in gametophytes. The gametophytes grow from haploid spores after sporic meiosis. The existence of a multicellular, haploid phase in the life cycle between meiosis and gametogenesis is also referred to as alternation of generations.

Gametogenesis in Animals

Animals produce gametes directly through meiosis in organs called gonads. Males and females of a species that reproduces sexually have different forms of gametogenesis:

- spermatogenesis (male)
- oogenesis (female).

Stages

However, before turning into gametogonia, the embryonic development of gametes is the same in males and females.

Common Path

Gametogonia are usually seen as the initial stage of gametogenesis. However, gametogonia are themselves successors of primordial germ cells.

During early embryonic development, primordial germ cells (PGCs) from the dorsal endoderm of the yolk sac migrate along the hindgut to the gonadal ridge.

They multiply by mitosis and once they have reached the gonadal ridge in the late embryonic stage, they are called gametogonia. Gametogonia are no longer the same between males and females.

Individual Path

From gametogonia, male and female gametes develop differently - males by spermatogenesis and females by oogenesis.

Gametogenesis in Gametangia

Fungi, algae and primitive plants form specialized haploid structures called gametangia where gametes are produced through mitosis. In some fungi, for example zygomycota, the gametangia are single cells on the end of hyphae and acting as gametes by fusing into a zygote.

More typically, gametangia are multicellular structures that differentiate into male and female organs:

- antheridium (male)
- archegonium (female).

Gametogenesis in Flowering Plants

In flowering plants, the male gamete is produced inside the pollen grain through the division of a generative cell into two sperm nuclei.

Depending on the species, this can occur while the pollen forms in the anther or after pollination and growth of the pollen tube. The female gamete is produced inside the embryo sac of the ovule.

Microgametogenesis

Microgametogenesis is the process in plant reproduction where a microgametophyte develops in a pollen grain to the three-celled stage of its development. In flowering plants it occurs with a microspore mother cell inside the anther of the plant.

When the microgametophyte is first formed inside the pollen grain four sets of fertile cells called sporogenous cells are apparent. These cells are surrounded by a wall of sterile cells called the tapetum, which supplies food to the cell and eventually becomes the cell wall for the pollen grain.

These sets of sporogenous cells eventually develop into diploid microspore mother cells. These microspore mother cells, also called microsporocytes, then undergo meiosis and become four microspore haploid cells.

These new microspore cells then undergo mitosis and form a tube cell and a generative cell. The generative cell then undergoes mitosis one more time to form two male gametes, also called sperm.

Fertilization

Fertilisation (also known as conception, fecundation and syngamy), is the fusion of gametes to produce a new organism. In animals, the process involves the fusion of an ovum with a sperm, which eventually leads to the development of an embryo.

Depending on the animal species, the process can occur within the body of the female in internal fertilisation, or outside in the case of external fertilisation.

The entire process of development of new individuals is called procreation, the act of species reproduction.

Fertilisation in Plants

Flowering Plants

After the carpel is pollinated, the pollen grain germinates in a response to a sugary fluid secreted by the mature stigma (mainly sucrose). From each pollen grain, a pollen tube grows out that attempts to travel to the ovary by creating a path through the female tissue. The vegetative (or tube) and generative nuclei of the pollen grain pass into its respective pollen tube.

After the pollen grain adheres to the stigma of the carpel (female reproductive structure) a pollen tube grows and penetrates the ovule through a tiny pore called a micropyle.

The pollen tube does not directly reach the ovary in a straight line. It travels near the skin of the style and curls to the bottom of the ovary, then near the receptacle, it breaks through the ovule through the micropyle (an opening in the ovule wall) and the pollen tube "bursts" into the embryo sac.

After being fertilised, the ovary starts to swell and will develop into the fruit. With multi-seeded fruits, multiple grains of pollen are necessary for syngamy with each ovule. The growth of the pollen tube is controlled by the vegetative (or tube) cytoplasm.

Hydrolytic enzymes are secreted by the pollen tube that digest the female tissue as the tube grows down the stigma and style; the digested tissue is used as a nutrient source for the pollen tube as it grows.

During pollen tube growth toward the ovary, the generative nucleus divides to produce two separate sperm nuclei (haploid number of chromosomes) - a growing pollen tube therefore contains three separate nuclei, two sperm and one tube.

The sperms are interconnected and dimorphic, the large one, in a number of plants, is also linked to the tube nucleus and the interconnected sperm and tube nucleuses form the "male germ unit". Double fertilisation is the process in angiosperms (flowering plants) in which two sperm nuclei from each pollen tube fertilise two cells in an ovary.

After the pollen tube reaches the ovary the pollen tube nucleus disintegrates and the two sperm cells are released into the ovary; one of the two sperm cells *fertilises* the egg cell (at the bottom of the ovule near the micropyle), forming a diploid (2n) zygote.

This is the point when fertilisation actually occurs. Note that pollination and fertilisation are two separate processes.

The other sperm cell fuses with two haploid polar nuclei (contained in the central cell) in the centre of the embryo sac (or ovule). The resulting cell is triploid (3n). This triploid cell divides through mitosis and forms the endosperm, a nutrient-rich tissue, inside the seed.

The two central cell maternal nuclei (polar nuclei) that contribute to the endosperm arise by mitosis from a single meiotic product. Therefore, maternal contribution to the genetic constitution of the triploid endosperm is different from that of the embryo.

Double fertilisation occurs only in angiosperm plants. One primitive species of flowering plant, *Nuphar polysepala*, has endosperm that is diploid, resulting from the fusion of a pollen nucleus with one, rather than two, maternal nuclei. It is believed that early in the development of angiosperm linages, there was a duplication in this mode of reproduction, producing seven-celled/eight-nucleate female gametophytes, and triploid endosperms with a 2:1 maternal to paternal genome ratio.

The process is easy to visualise if one looks at maize silk, which is the female flower of corn. Pollen from the tassel (the male flower) falls on the sticky external portion of the silk, and then pollen tubes grow down the silk to the attached ovule. The dried silk remains inside the husk of the ear as the seeds mature; if one carefully removes the husk, the floral structures may be seen.

In many plants, the development of the flesh of the fruit is proportional to the percentage of fertilised ovules. For example, with watermelon, about a thousand grains of pollen must be delivered and spread evenly on the three lobes of the stigma to make a normal sized and shaped fruit.

Fertilisation in Animals

The mechanics behind fertilisation has been studied extensively in sea urchins and mice. This research addresses the question of how the sperm and the appropriate egg find each other and the question of how only one sperm gets into the egg and delivers its contents. There are three steps to fertilisation that ensure species-specificity:

1. Chemotaxis
2. Sperm activation/acrosomal reaction
3. Sperm/egg adhesion.

Internal vs. External

Consideration as to whether an animal (more specifically a vertebrate) uses internal or external fertilisation is often dependent on the method of birth. Oviparous animals laying eggs with thick calcium shells, such as chickens, or thick leathery shells generally reproduce via internal fertilisation so that the sperm fertilise the egg without having to pass through the thick, protective, tertiary layer of the egg. Ovoviviparous and euviviparous animals also use internal fertilisation.

It is important to note that although some organisms reproduce via amplexus, they may still use internal fertilisation, as with some salamanders. Advantages to internal fertilisation include: minimal waste of gametes; greater chance of individual egg fertilisation, relatively "longer" time period of egg protection, and selective fertilisation; many females have the ability to store sperm for extended periods of time and can fertilise their eggs at their own desire.

Oviparous animals producing eggs with thin tertiary membranes or no membranes at all, on the other hand, use external fertilisation methods. Advantages to external fertilisation include: minimal contact and transmission of bodily fluids; decreasing the risk of disease transmission, and greater genetic variation (especially during broadcast spawning external fertilisation methods).

Sea Urchins

Chemotaxis was discovered as the method by which sperm find the eggs. This chemotaxis is an example of a ligand/receptor interaction. Resact is a 14 amino acid peptide purified from the jelly coat of A. punctulata that attracts the migration of sperm. After finding the egg, the sperm gets through the jelly coat through a process called sperm activation. In another ligand/receptor interaction, an oligosaccharide component of the egg binds and activates a receptor on the sperm and causes the acrosomal reaction.

The acrosomal vesicles of the sperm fuse with the plasma membrane and are released. In this process, molecules bound to the acrosomal vesicle membrane, such as bindin, are exposed on the surface of the sperm. These contents digest the jelly coat and eventually the vitelline membrane. In addition to the release of acrosomal vesicles, there is explosive polymerization of actin to form a thin spike at the head of the sperm called the acrosomal process.

The sperm binds to the egg through another ligand reaction between receptors on the vitelline membrane. The sperm surface protein bindin, binds to a receptor on the vitelline membrane identified as ERB1.

Fusion of the plasma membranes of the sperm and egg are likely mediated by bindin. At the site of contact, fusion causes the formation of a fertilisation cone.

Mammals

Usually mammals rely on internal fertilisation through copulation. After a male ejaculates, a large number of sperm cells move to the upper vagina (via contractions from the vagina) through the cervix and across the length of the uterus toward the ovum.

The capacitated spermatozoon and the oocyte meet and interact in the *ampulla* of the fallopian tube. It is probable that chemotaxis is involved in guiding the sperm to the egg, but the mechanism has yet to be worked out.

However, demonstration of formyl peptide receptors (60.000 receptor/cell; higher binding capacity in the tail region) in the surface membrane of human sperms strongly supports, that - besides specific chemoattractant substances i.e. resact - professional chemoattractant ligands like formyl Met-Leu-Phe (fMLF) have also the ability to induce migration of sperm.

The zona pellucida of the egg binds with the sperm. In contrast to sea urchins, the sperm binds to the egg before the acrosmal reaction. The zona pellucida is a thick layer of extracellular matrix that surrounds the egg and is similar to the role of the vitelline membrane in sea urchins. A glycoprotein in the zona pellucida, ZP3 was discovered to be responsible for egg/sperm adhesion *in mice.* The receptor galactosyltransferase (GalT) binds to the N-acetylglucosamine residues on the ZP3 and is important for binding with the sperm and activating the acrosome reaction. ZP3 is sufficient for sperm/egg binding but not necessary.

There are two additional sperm receptors: a 250kD protein that binds to an oviduct secreted protein and SED1 which binds independently to the zona.

After the acrosome reaction, it is believed that the sperm remains bound to the zona pellucida through exposed ZP2 receptors. These receptors are unknown in mice but have been identified in guinea pigs.

In mammals, binding of the spermatozoon to the GalT initiates the acrosome reaction. This process releases the enzyme hyaluronidase, which digests the matrix of hyaluronic acid in the vestments surrounding the oocyte.

Fusion between the oocyte plasma membranes and sperm follows, allowing the entry of the sperm nucleus, centriole and flagellum, but not the mitochondria, into the oocyte. The fusion is likely mediated by the protein CD9 in mice (the binding homolog). The egg "activates" itself upon fusing with a single sperm cell, thereby changing its cell membrane to prevent fusion with other sperm.

This process ultimately leads to the formation of a diploid cell called a zygote. The zygote begins to divide and form a blastocyst and when it reaches the uterus, it performs implantation in the endometrium.

At this point the female's pregnancy has begun. If the embryo implants in any tissue other than the uterine wall, an ectopic pregnancy results, which can be fatal to the mother.

In some animals (e.g. rabbits) the act of coitus induces ovulation by stimulating release of the pituitary hormone gonadotropin. This greatly increases the probability that coitus will result in pregnancy.

Humans

Human fertilization is the union of a human egg and sperm, usually occurring in the ampulla of the uterine tube. It is also the initiation of prenatal development. Scientists discovered the dynamics of human fertilisation in the nineteenth century.

The process of fertilization involves a sperm fusing with an ovum—usually following ejaculation during sexual intercourse. It is possible, but less probable, for fertilization to occur without sexual intercourse, artificial insemination, or In vitro fertilisation.

Upon encountering the ovum, the acrosome of the sperm produces enzymes which allow it to burrow through the outer jelly coat of the egg. The sperm plasma then fuses with the egg's plasma membrane, the sperm head disconnects from its flagellum and the egg travels down the Fallopian tube to reach the uterus. In vitro fertilisation (IVF) is a process by which egg cells are fertilized by sperm outside the womb, in vitro.

Anatomy

Corona Radiata

The egg and the sperm bind through the corona radiata, a layer of follicle cells on the outside of the secondary oocyte. Fertilization occurs when the nuclei of a sperm and an egg fuse. The successful fusion of gametes form a new organism.

Acrosome Reaction

The acrosome reaction must occur to mobilise enzymes within the head of the spermatozoon to degrade the zona pellucida. example:

Zona Pellucida

After binding to the corona radiata the sperm reaches the zona pellucida, which is an extra-cellular matrix of glycoproteins. A special complementary molecule on the surface of the sperm head binds to a ZP2 glycoprotein in the zona pellucida.

This binding triggers the acrosome to burst, releasing enzymes that help the sperm get through the zona pellucida. Some sperm cells consume their acrosome prematurely on the surface of the egg cell, facilitating the penetration by other sperm cells.

As a population, sperm cells have on average 50% genome similarity so the premature acrosomal reactions aid fertilization by a member of the same cohort.

It may be regarded as a mechanism of kin selection. Recent studies have shown that the egg is not passive during this process.

Cortical Reaction

Once the sperm cells find their way past the zona pellucida, the cortical reaction occurs: cortical granules inside the secondary oocyte fuse with the plasma membrane of the cell, causing enzymes inside these granules to be expelled by exocytosis to the zona pellucida.

This in turn causes the glyco-proteins in the zona pellucida to cross-link with each other, making the whole matrix hard and impermeable to sperm.

This prevents fertilization of an egg by more than one sperm.

Fusion

After the sperm enters the cytoplasm of the oocyte, the cortical reaction takes place, preventing other sperm from fertilizing the same egg.

The oocyte now undergoes its second meiotic division producing the haploid ovum and releasing a polar body. The sperm nucleus then fuses with the ovum, enabling fusion of their genetic material.

Cell Membranes

The cell membranes of the secondary oocyte and sperm fuse.

Transformations

In preparation for the fusion of their genetic material both the oocyte and the sperm undergo transformations as a reaction to the fusion of cell membranes. The oocyte completes its second meiotic division. This results in a mature ovum. The nucleus of the oocyte is called a pronucleus in this process, to distinguish it from the nuclei that are the result of fertilization.

The sperm's tail and mitochondria degenerate with the formation of the male pronucleus. This is why all mitochondria in humans are of maternal origin

Replication

The pronuclei migrate toward the centre of the oocyte, rapidly replicating their DNA as they do so to prepare the embryo for its first mitotic division.

Mitosis

The male and female pronuclei don't fuse, although their genetic material do. Instead, their membranes dissolve, leaving no barriers between the male and female chromosomes. During this dissolution, a mitotic spindle forms between them.

The spindle captures the chromosomes before they disperse in the egg cytoplasm. Upon subsequently undergoing mitosis (which includes pulling of chromatids towards centrioles in anaphase) the cell gathers genetic material from the male and female together. Thus, the first mitosis of the union of sperm and oocyte is the actual fusion of their chromosomes.

Each of the two daughter cells resulting from that mitosis has one replica of each chromatid that was replicated in the previous stage. Thus, they are genetically identical.

Diseases

Various disorders can arise from defects in the fertilization process.

- Polyspermy results from multiple sperm fertilizing an egg.

However, some researchers have found that in rare pairs of fraternal twins, their origin might have been from the fertilization of one egg cell from the mother and two sperm cells from the father. This possibility has been investigated by computer simulations of the fertilization process.

The term *conception* commonly refers to fertilisation, the successful fusion of gametes to form a new organism. 'Conception' is used by some to refer to implantation and is thus a subject of semantic arguments about the beginning of pregnancy, typically in the context of the abortion debate.

Gastrulation, which occurs around 16 days after fertilisation, is the point in development when the implanted blastocyst develops three germ layers, the endoderm, the ectoderm and the mesoderm. It is at this point that the genetic code of the father becomes fully involved in the development of the embryo.

Until this point in development, twinning is possible. Additionally, interspecies hybrids survive only until gastrulation, and have no chance of development afterward.

However this stance is not entirely accepted as some human developmental biology literature refers to the "conceptus" and such medical literature refers to the "products of conception" as the post-implantation embryo and its surrounding membranes. The term "conception" is not usually used in scientific literature because of its variable definition and connotation.

Fertilisation and Genetic Recombination

Meiosis results in a random segregation of the genes contributed from each parent. Each parent organism generally has the same genetic make-up, but differs for a fraction of their genes.

Therefore, each gamete produced by a person will be genetically different from the others from that person, as well as from the gametes produced by another person.

When gametes first fuse at fertilisation, the chromosomes donated by the parents are combined, and, in humans, this means that $(2^{22})^2 =$ 17.6x10 chromosomally different zygotes are possible for the non-sex chromosomes, even assuming no chromosomal crossover.

If crossover occurs once, then on average $(4^{22})^2$ = 309x10 genetically different zygotes are possible for every couple, not considering that crossover events can take place at most points along each chromosome. The X and Y chromosomes do not undergo crossover events, so are excluded from the calculation. Note that the mitochondrial DNA is only inherited from the maternal parent.

Parthenogenesis

Another method of fertilisation occurs among animals that normally reproduce sexually, through parthenogenesis: when the gamete of a female is not fertilised by a male, yet produces viable and unique offspring that are not clones. Only DNA from the mother is inherited, but it is not identical to her. Normal eggs of the mother become fertilised, without sperm, and development proceeds normally. This occurs naturally in several species and may be induced in others through a chemical or electrical stimulus. In 2004, Japanese researchers led by Tomohiro Kono succeeded after 457 attempts to merge the ova of two mice, the result of which developed normally into a mouse. This was achieved by blocking certain proteins that would normally prevent the possibility.

Chapter 2

The Chemical Basis of Life: Atoms and Molecules

All matter is made up of tiny particles called *atoms*. Atoms are the basic building blocks of all mater, including life. They cannot be broken down by any known means. There are only about 100 kinds of atoms but they can be chemically combined to make thousands of different kinds of substances.

The substances are called *elements* if they are made up of only one kind of atom. Hydrogen gas is an example of an element because there are two atoms of Hydrogen joined together.

Elements are pure sunstances that cannot be broken down by any oridnay means. Unlike solutions, elements are made up of only one kind of atom with the same atomic number.

For example, the element Nitrogen consists entirely of Nitrogen atoms. There are over 100 different kinds of elements known today. These elements are organized into different families and different periods in the periodic table.

They can be in the form of any state, solid, liquid, or a gas. Compounds are made up of two or more different *elements* put together in definite proportions. For example, water is a compound because it contains two atoms of Hydrogen for every one atom of Oxygen.

A molecule is like a compound in that it is two or more atoms combined, but each atom when combined has to act as a single particle. If a sodium atom passes an electron to a chlorine atom then the chlorine atom has 17 protons and 18 electrons. An atom that has an excess charge is called an ion.

Compounds

Most substances are compounds. A compound is made up of two or more different kinds of atoms combined in definite proportions. Compounds can be separated into the individual atoms that compose their composition, unlike elements that cannot. Compounds can be seperated only by a chemical change that involves heat or electrcity. For example, Water can be separated int Hydrogen and Oxygen by applying a sufficient DC voltage.

Ionic Bonding

Ionic bonding is the bond between two atoms formed as a result of the transfer of electrons from one atom to the other. This results in the formation of positive and negative ions. IONS - Any atom which has an excess charge is called an ion, and the force of attraction between two ions is called an ionic bond. Example: When a sodium atom gives up an electron, it will have 11 protons and 10 electrons. This creates a sodium ion which has an excess positive charge of 1 unit. Relate the concepts of solutions, mixtures, acids and bases to the composition and environment of cells.

A solution is a type of mixture that has the same composition throughtout, and appears to be all one substance. In other words, it's a homogeneous mixture that can be in the form of a solid, liquid or gas. The make up of cells consist of many types of mixtures, including solutions.

Many of them are in the form of a liquid. A solution is formed by mixing one or more substances into another substance. In any solution, there is a solute and a solvent. The solvent is generally a liquid, and it is the substance in which the solute is dissolved in. A mixture is a word used to describe a blend of different substances. A mixture is made up of two or more substances, in which they can all be visually seperated and distinguished from one another.

Solutions

A solution is any homogeneous mixture, and may include solids, liquids, as well as gases. Most solutions important to living cells are thought of as liquids. The liquid substance that makes up the bulk of a solution is known as a solvent. Substances that dissolve in the solvent are called solutes (SAHL yoots). A solute can either be a solid, liquid, or gas before they are dissolved in the solvent. The most common solutions have water as the solvent.

Mixtures

A mixture is a combination of substances that are physically mixed without forming new chemical bonds. The substances in a mixture may be present in any proportions. The proportions can change as one substance may be added to, or removed from the mixture, but the different substances in the mixture retain their usual properties.

Acids

An acid is any compound that produces hydrogen ions in a solution. All acids consist of molecules in which contain hydrogen. The molecules are covalently bonded to another atom or group of atoms. When these molecules dissociate in a solution, such as water, the hydrogen breaks loose as an hydrogen ion. The left over molecules form what is called a negative ion. Hydrochloric acid is an example of an acid, which is important to life activities. This is because the level of acid in the environment of the cell affects the actions of enzymes and therefore affects the functioning of the cell.

Base

A base is any compound that produces excess hydroxide ions when it is dissolved in a solution. When in a dry state many bases are ionic compounds. A perfect example of a base is sodium hydroxide. Sodium hydroxide is a solid compound, which is made up of sodium and hydroxide ions. When a base is dissolved in water, it dissociates (separates) into its ions:

$$NaOH \longrightarrow Na^+ + OH^-$$

An acid is any compound that produces hydrogen ions in solution. All acids consist of molecules that contain hydrogen covalently bonded to another atom or group of atoms.

A base is a compound that produces hydroxide(hy DRAHK syd) ions (OH-) when disolved in water. In the dry slate, many bases are ionic compounds. Sodium hydroxide (NaOH) is one example. This is a solid compound made of sodium and hydroxide ions. When dissolved in water it separates in to it's ions. NaOH then dissociates to form Na+ and OH- ions.

When acid and a base react their hydrogen and hydroxide combine to form water molecules. This reaction removes these two ions from the solution. that is, they are no longer present as separate ions. However, the negative ions of the acid and the positive ions of the base are still present. Identify the importance of water to life by relating its functions as a transport and dissolving medium.

Water is very important to life. Some scientists beleived that life probaly evolved in water. All living cells are made up of 70-95% water and the water covers 75% of the earths surface. Water is present in all three physical states of mater - solid, liquid, and gas.

Water gets its properties from the fact that it is a polar covalent molecule, and therefore forms hydrogen bonds that connects one water molecule to as many as four other water molecules. This results in a property known as cohesion.

Water also has adhesive properties to ionic substances as well as other charged substances such as clay. These properties result in many of water's important properties which include its remarkable ability to dissolve a vast variety of covalent and ionic substances, its high heat capacity, capillarity, surface tension, high heat of vaporization, and the fact that water floats when it freezes. All of these properties of water are related to life on Earth.

Water is considered the universal solvent of life. It acts as a solvent which transports molecules and ions from one area of a cell to another. It also acts as a reactant in chemical reactions that occur in the cell. Without water, molecules and ions would be unable to diffuse through the membrane that surrounds the cell, and gases such as carbon dioxide and oxygen would be unable to pass through plasma membranes.

Describe the Characteristics of Organic Compounds

Organic compounds contain carbon atoms that are linked by covalent bonds to form molecular chains, branches, rings, or sheets. Because there are 4 electrons in the outer energy level of the carbon atom, carbon can combine with other carbon atoms, as well as to the many other biologically important elements such as hydrogen, nitrogen, oxygen, sodium, magnesium, phosphorus, sulfur, chlorine, potassium, calcium, and iron.

In addition to carbon, the vast majority of organic compounds also contain hydrogen. Oxides, carbonates and cyanides also contain carbon atoms but are not considered to be organic. Such compounds as carbon monoxide, carbon dioxide, calcium carbonate, and cyanide are considered to be inorganic, even though they contain carbon atoms.

Make an analysis of carbohydrates, lipids, proteins, and nucleic acids in terms of: the atoms that compose these compounds, the function these compounds serve in the body, the food sources of each compound, the sub-units that chemically combine to make these compounds, and examples of each group.

The first compound is carbohydrates. Carbohydrates are sugars and starches. Carbohydrates are found in breads, cereals, potatoes, pastas etc. Lipids is the next compound. Lipids are fats, oils and waxes. Lipids are found in mainly fried foods such as pizza, fries, etc.

Protein is the next compound. Proteins are compounds that contain nitrogen as well as carbon, hydrogen and oxygen. Proteins are found in beans, fish, and meats. The final compound is nucleic acids. Nucleic acids contain phosphorous and nitrogen along with carbon, hydrogen and oxygen. Nucleic acids are mainly found in meats

Carbohydrates are compunds composed of carbon, hydrogen and oxygen. The name carbohydrates comes from the fact that they are formed from the union of carbon dioxide and water during the process of photosynthesis.

The ratio of hydrogen to oxygen in a carbohydrate is always two to one - two hydrogen atoms for every oxygen atom. Carbohydrates are normally foun in the breads and cereals as well as fruits and vegetables.

The simplest carbohydrates are just simple sugars (called monosaccharides). These sugars are very important to the boy, because they contain large amounts of energy. This energy can be released in the presence of oxygen by breaking down the sugars into carbon dioxide and water.

Lipids include subtances commonly known as fats, waxes and oils. Like carbohydrates, lipids are composed of carbon, hydrogen and oxygen, except there is less oxygen in lipids, than in carbohydrates. Lipids help make up parts of the cell structure (such as the cell membranes) and are used as a reserve for energy in an organism. Lipids provide twice as much energy per gram as the same amount of carbohydrate. Lipids are found in fatty meats such as bacon, as well as products such as whole milk, cream and butter, and from the oils of many plants such as olives, corn, and peanuts.

Proteins are compuounds containing nitrogen, carbon, hydrogen & oxygen. Some proteins also contain sulfur and phosphorus. The number of proteins in your body is virtually countless. Proteins are found everywhere, in cartilage, bones and muscles, hormones, antibodies, and even enzymes. Proteins are notmally found in meats, poltry, fish and some plants such as beans, soy beans, and legumes.

Nucleic Acids are organic compounds that contain Phosphorus, and Nitrogen, as well as Hydrogen, Oxygen and Carbon. Ther are only two known types of nucleic acids, and they are DNA (Deoxyribonucleic

Acid) and RNA (Ribonucleic Acid). These two substances were first found in the Nucleus of a Cell. DNA (which is hereditary) and RNA direct and control the activities and structural devolpment of all the cells of an organism.

Specify the Importance of Nucleic Acids

Nucleic Acids are compounds that contain phosphorous and nitrogen in addition to carbon. There are two types of nucleic acids that are formed by the nucleolus located within the nucleus of cells. DNA (deoxyribonucleic acid) and RNA (ribonucleic acid) carry the genetic code that controls the manufacture or synthesis of proteins. These proteins serve as building blocks of cells, as hormones that regulate cellular function, as antibodies that fight infection, and as enzymes that regulate chemical reactions within the cell.

Nucleic acids are a group of organic substances found in the chromosomes of living cells and viruses. They play a central role in the storage and replication of hereditary information and in the expression of this information through protein synthesis. In most organisms, nucleic acids occur in combination with proteins; the combined substances are called nucleoproteins. Nucleic acid molecules are complex chains of varying length. The two chief types of nucleic acids are DNA (deoxyribonucleic acid), which carries the hereditary information from generation to generation, and RNA (ribonucleic acid), which delivers the instructions coded in this information to the cell's protein manufacturing sites (ribosomes).

Outline the function of minerals such as calcium, phosphorus, potassium, and sodium and describe why these substances are essential.

Minerals are chemical elements that organisms need for normal functioning. The most important minerals are iron, calcium, phosphorus, and iodine.

Calcium is the most abundant mineral in the human body, and for that matter, the body of any boned animal, as it is the structural element of bones. Weather our body is deficient in calcium depends less on whether we have a sufficient dietary intake, but rather on how well we absorb it. An important factor in calcium absorption is Vitamin D.

The healthy body can produce this vitamin itself for as long as it gets some exposure to sunlight every day. Calcium absorption depends on hormones. Women beyond menopause are under increased risk of developing osteoporosis(bone loss), not because they would lacking in dietary calcium but because they lack the estrogen needed for proper absorption. Trace minerals such as boron also play a part in calcium absorption. The best sources of dietary calcium are milk and milk products, as well as soy and green leafy vegetables. Recommended daily amounts for dietary calcium are about 700 milligram for a healthy adult, and about a gram for adolescents.

Phosphates are essential to the energy-transfer reactions necessary to sustain life processes. Another name for this is adenosine triphospate or ATP which is highly significant to all cells in your body which is involved in every metabolic and photosynthetic reaction.

In nature phosphorus occurs only in compounds called phosphates. Plants get phosphorus from soil, and use it in photosynthesis. People and animals take in phosphorus by eating plants and different types of foods.

Distinguish between carbohydrates (sugars and starches), lipids and proteins by using the appropriate tests to identify these nutrients. List each test and how it is conducted.

Tests for lipids: There are two tests that can be used to test for the presence of lipids:

1. Sudan IV : Sudna IV is a special biological stain that is soluble in nonpolar solvents such as lipids and turn from a pink to red colour. Polar compounds will not take up the pink or red colour. For example, if you take a test tube of water and vegetable oil (a lipid) and mix them together and then add a small amount of Sudan IV and shake the mixture, only the oil will turn red or pink not the water.
2. The second test is often called the translucence test for lipid. This is the test that uses unglazed brown paper (same type of brown paper you find in a brown paper bag like the ones from the Co-op used to rap bottles). If lipid is present, the lipid is absorbed into the paper and causes it to appear translucent which means that light is able to pass through and it looks like a greesy spot.

Tests for Protein: The Biuret Test is often used for protein. As you will learn soon, protein is made of a chain of amino acids that are bonded together by what are known as peptide bonds. The Biuret Test uses Biuret Reagent (a special solution) that is able to react with the peptide bonds of the amino acid chain that makes up the protein. This causes a colour change from blue indicating no protein, to pink (small amount of protein), to violet (larger amount of protein), to purple (largest amount of protein).

An example to use is to test a known sample of protein such as egg white or gelatin. Adding the Biuret reagent to either of these food materials will cause a purple colour. Adding Biuret Reagent to water would result in blue colour.

Tests for carbohydrate: Benedicts Reagent is used for single sugars (called monosaccharides). Presence of monosaccharides causes the Benedict's solution to change to various colours when heated. The colour change depends on the concentration of the sugar present. Benedict's Reagent is made from Copper Sulfate and therefore contains Cu^{2+} ions. In the presence of simple sugars (known as reducing sugars) the copper is changed to Cu^{1+} ions. The reaction depends on the concentration of the sugar present. Here is a list of colour changes:

- Blue means no sugar present.
- Light green means 0.5% to 1.0% sugar present.
- Green to Yellow means 1.0% to 1.5% sugar present.
- Orange means 1.5% to 2.0% sugar present.
- Red to red brown means greater than 2.0% sugar present.

Iodine is the test for starch. Mixing iodine and starch causes the formation of a starch-iodine complex that is blue-black in colour. If you place iodine on a sample of food and it remains an amber or light brown colour there is no starch present, but if there is starch present, then the colour will change to a dark blue-black colour.

Explain what vitamins are and describe in general terms the vital body functions of vitamins A, B, C, D, E, and K.

Vitamins are organic molecules required in the diet in very small amounts. Vitamins are substances that are normally made by plants and then ingested by consumers. The vitamins are important because they serve as coenzymes which help the enzymes function more effectively.

If the body is unable to supply those necessary enzymes in the proper quantities at the proper time, the vitamins and minerals become inert materials and pass unused through your body. Vitamins function directly as coenzymes, or are converted into coenzymes within the cell. Many important biological reactions need vitamins.

Vitamin K and some of the B vitamins are essential for blood clotting. This is because Vitmain K is needed for the symthesis of prothrombin, one of the plasma proteins that takes part in the series of reactions leading to the formation of a blood clott.

When large doses of anitbiotics are given to overcome an infection, they can destroy the intestinal bacteria, and a Vitamin K deficiency may result. This is because Vitamin K is produced as a product of the metabolic activities of these bacteria.

Vitamin A

Vitamin A is a fat-soluble vitamin, which is structurally related to carotene. This is because carotene is converted into vitamin A in the liver. It is composed of compounds called retinoids and carotenoids, and is measured in Retinol Equivalents (RE's) which measure the vitamin A activity of food. There are many vital body functions in which vitamin A is used, some of these are maintaining the protective linings of the lungs, intestine, urinary tract, and other organs, as well as maintaining good vision, and promoting proper bone growth and tooth development. Too little vitamin

A leads to a conditon called night blindness. Night blind people have trouble seeing in dim light. Both black-and-white and colour vision make use of the light-sensitive pigment retinal, which is made from Vitamin A. Others include helping in the formation and maintenance of healthy skin and hair, healthy mucus membranes, fortifying the immune system, reproduction, and hormone synthesis and regulation. Also, Vitamin A is used to treat minor skin disorders, it is important for the human metabolism, and is thought to help prevent the development of cancer. Sources of vitamin A include things such as some diary products, eggs, and liver.

Some sources of carotene are orange fruits and vegetables, as well as dark, leafy greens. The absence of Vitamin A in ones diet can cause loss in weight, eye diseases (including night blindness), kidney stones, rashes, depression, frequent infections of the respiratory system, bladder, or digestive systems, plus many other problems.

Vitamin B1 (Thiamine)

Vitamin B1 is a water-soluble vitamin which is important for the production of energy and metabolism of sugar. It is also essential for supporting normal appetite, nervous system functions, muscles and heart, as well as aiding in the digestion of carbohydrates, and promoting growth and good muscle tone. Good sources of vitamin B1 are some dairy products, nuts, meats, whole grains, yeast eggs, and germ of cereals. The lack of vitamin B1 in ones diet may cause the loss of appetite, loss of weight, nausea, anxiety, irritability, insomnia, depression, weakness and fatigue. Other problems that may occur are heart and muscle problems, such as enlarged heart, abnormal heart rhythms, heart failure, muscle wasting, aches and pains in muscles, difficulty walking, loss of reflexes, and paralysis.

Vitamin B2 (Riboflavin)

Vitamin B2 is a water-soluble vitamin which is found to release energy from protein, fat, and carbohydrates, and is the precursor of flavoproteins, flavin-adenine dinucleotide and flavin mononucleotide. This vitamin is part of a coenzyme used in energy metabolism, it alleviates eye fatigue, helps maintain cell respiration, healthy eyes, hair, skin, and nails, aids in the formation of antibodies and red blood cells, and promotes general health. Good sources of vitamin B2 are some dairy products, liver, the white of an egg, meats, whole grains, cereals, almonds, and green vegetables. Some problems that are caused by the dietary lack of vitamin B2 are magenta tongue, rashes, cracks and sores in or around the mouth, itching, burning, bloodshot eyes, digestive disturbances, trembling, sluggishness, as well as dry, scaly skin.

Vitamin B3 (Niacin)

Vitamin B3 is a water-soluble vitamin, it's precursor is tryptophan, and it is involved in the oxidative release of energy from food. It is necessary for the metabolism of carbohydrates, is part of a coenzyme used in energy metabolism, helps improve circulation, supports the health of the nervous system, digestive system, and the skin, as well as protects the skin.

Other vital body functions of vitamin B3 are preventing pellagra, and reducing high blood pressure, and help lower cholesterol levels. Dietary sources of vitamin B3 include milk, eggs, meat, whole grains and cereals, nuts, and all protein-containing foods.

Insufficient amounts of vitamin B3 in one's diet may result in pellagra, irritability, nervousness, loss of appetite, dizziness, mental confusion, flaky skin rash on sun exposed areas, and delirium. Others include depression, fatigue, headaches, muscular weakness, insomnia, and black, smooth tongue.

Vitamin B5 (Pantothenic Acid)

Vitamin B5 is a water-soluble vitamin which is essential for the metabolism of food. Some vital body functions of vitamin B5 include metabolism of proteins, carbohydrates, and fats, and the production of steroid hormones and other important chemicals, helps fight depression.

As well as supporting the normal functioning of the gastrointestinal tract, and it is required for the production of cholesterol, bile, red blood cells, and antibodies. Good dietary sources of vitamin B5 are whole grains, beans, liver, eggs, fish, milk and milk products, yeast, lean beef, some potatoes. Others include broccoli, and other vegetables in the cabbage family. Deficiency of vitamin B5 causes problems such as abdominal distress, burning sensation in the heels, and sleep problems.

Vitamin B6 (Pyridoxine)

Vitamin B6 is a water-soluble vitamin, and is related compounds pyridoxamine, and pyridoxal. This vitamin is necessary for the synthesis and breakdown of amino acids, is used for the metabolism of protein, aids in the formation of red blood cells, helps convert tryptophan to niacin, and for maintaining the central nervous system. It also helps maintain a proper balance of sodium and phosphorous in the body, aids in the removal of excess fluid in premenstrual women, promotes healthy skin, and reduces muscle spasms, leg cramps, nausea, as well as hand numbness and stiffness.

Sources of vitamin B6 include bananas, fish, shellfish, meat, poultry, nuts, fruits, whole grains, vegetables, yeast, wheat germ, maize, and rice husks. The chance of deficiency is increased by smoking, and can cause problems such as anaemia, smooth tongue, insomnia, irritability, muscle twitching, convulsions, rashes, kidney stones, greasy dermatitis. In addition, other resulting problems are arm and leg cramps, loss of hair, muscular weakness, slow learning, mouth disorders, water retention, nervousness, and skin eruptions.

Vitamin B12 (Cyanocobalamin)

Vitamin B12 is a water-soluble vitamin, and is involved in the biosynthesis of methyl groups of choline, and methionine. Some vital body functions of vitamin B12 are being part of a coenzyme used in new cell synthesis, help maintain and protect the nerve cells and system, helps in the formation and regeneration of red blood cells, and helps prevent anaemia.

Others include being used for genetic materials, it is necessary for carbohydrate, fat and protein metabolism, promotes growth in children, increases energy, and is needed for calcium absorption. Significant sources of this vitamin are milk, cheese, eggs, liver and meat.

A shortage of vitamin B12 in one's diet can lead to anaemia, poor appetite, growth failure in children, fatigue, depression, lack of balance, smooth tongue, and skin hypersensitivity. In addition, other problems are brain damage, paralysis, nervous system degeneration, as well as degeneration of the spinal cord.

Define Enzymes and Explain the Make-up and Importance of Enzymes

Enzymes (EN zymz) are protein substances that are necessary for most of the chemical reactions that occur in living cells. Enzymes speed up chemical reactions. Enzymes go into a chemical reaction only temporarily. They are not changed by this reaction so they may be used over and over again for the same chemical step with other molecules. In this way, enzymes are organic catalysts. The substance that an enzyme acts upon is called it's substrate.

Enzymes are almost invisible Protein molecules which are contained in the food we eat and are produced by our bodies. Enzymes are necessary for most of the chemical reations that occur in living cells. All enzymes in an organism are made by the cells of the organism. There are three different types of enzymes which are Metabolic, Digestive, and Food. Enzymes are important because they purify blood, strengthen the immune system, break down fats, lower cholesterol, enhance mental capacity, cleanse the colon, enhance sleep, and improve aging skin.

All living cells make enzymes, but enzymes are not alive. Enzyme molecules function by altering other molecules. Enzymes combine with the altered molecules to form a complex molecule structure in which chemical reactions take place. The enzyme, which remains unchanged, then seprates from the product of the reaction. The human body has thousands of kinds of enzymes. Each kind does it's own job. Without these people could not breath, see, move, or digest food. Photosynthesis in plants also depends on enzymes.

There is a major difference between chemical reactions in nonliving things and the reactions in living things. Consider, for example, the burning of gasoline in an automobile engine. The gasoline vapor is admitted to the engine cylinder. It is ignited by a spark. The vapor burns in a fraction of a second. In fact, the burning is so rapid that it produces a small explosion, which helps to drive the engine. How Enzymes work is that the ability to act as a catalysts depends on their shape. Somewhere on the surface of each enzyme there is a region called the active site.

An Enzyme may also cause 2 molecules to join. In this case, there are two substrates. Each fits into the active site in such a way that they are brought into close contact. This enables bonds to form between the two substrate molecules. There is a theory to explain how enzyme and substrate fit together at an active site. It is called the lock-and-key model. Just as the notched surface of a key can open only one lock, the shape of the active site of an enzyme only fits certian substrates. Thus, each enzyme can catalyze a reaction only of those substrates.

The more modern view of enzyme function is known as the induced fit model which suggests that the enzyme changes shape just slightly to better fit the substrate.

Enzymes are proteins which act as biological catalysts. A catalyst enables a chemical reaction to take place at lower temperatures by lowering the activation energy required for the reaction to occur.

This is important to living cells since the enzyme enables the metabolic reactions to occur at normal cellular temperatures. Many enzymes must be attached to certain non-protein molecules in order to function. Some molecules are called cofactors, such as copper, iron, and magnesium. Others are called coenzymes.

Many coenzymes consist of vitamins, especially B vitamins. If a persons diet lacks the right amount of these vitamins, the enzymes cannot function properly, and serious problems may occur within the body.

Molecules of Life: Biomolecules

A biomolecule is any organic molecule that is produced by a living organism, including large polymeric molecules such as proteins, polysaccharides, and nucleic acids as well as small molecules such as primary metabolites, secondary metabolites, and natural products. A more general name for this class of molecules is a biogenic substance.

As organic molecules, biomolecules consist primarily of carbon and hydrogen, nitrogen, and oxygen, and, to a smaller extent, phosphorus and sulfur. Other elements sometimes are incorporated but are much less common.

Types of Biomolecules

A diverse range of biomolecules exist, including:

- Small molecules:

In the fields of pharmacology and biochemistry, a small molecule is a low molecular weight organic compound which is by definition not a polymer. The term small molecule, especially within the field of pharmacology, is usually restricted to a molecule that also binds with high affinity to a biopolymer such as protein, nucleic acid, or polysaccharide and in addition alters the activity or function of the biopolymer. The upper molecular weight limit for a small molecule is approximately 800 Daltons which allows for the possibility to rapidly diffuse across cell membranes so that they can reach intracellular sites of action. In addition, this molecular weight cutoff is a necessary but insufficient condition for oral bioavailability.

Small molecules can have a variety of biological functions, serving as cell signaling molecules, as tools in molecular biology, as drugs in medicine, as pesticides in farming, and in many other roles. These compounds can be natural (such as secondary metabolites) or artificial (such as antiviral drugs); they may have a beneficial effect against a disease (such as drugs) or may be detrimental (such as teratogens and carcinogens).

Biopolymers such as nucleic acids, proteins, and polysaccharides (such as starch or cellulose) are not small molecules, although their constituent monomers — ribo- or deoxyribonucleotides, amino acids, and monosaccharides, respectively — are often considered to be. Very small oligomers are also usually considered small molecules, such as dinucleotides, peptides such as the antioxidant glutathione, and disaccharides such as sucrose.

Drugs

Most drugs are small molecules, although some drugs can be proteins, e.g. insulin. Many proteins are degraded if administered orally and most often cannot cross the cell membranes. Small molecules are more likely to be absorbed, although some of them are only absorbed after oral administration if given as prodrugs. Many dietary supplements are small molecules (but not herb extracts, such as ginkgo).

Primary and Secondary Metabolites

For organisms to produce small molecules they need one or more specialized enzymes (to create and destroy), which as a result are not that varied in vertebrates (recent and small + slow population size), but very common in soil bacteria (such as streptomyces) and fungi, which in particular secrete antibiotics.

Plants also have several secondary metabolites, which play a role in cell signalling, pigmentation or in defence, several of which have also been used as drugs (medical and recreational):

- Alkaloids
- Glycosides
- Lipids
- Flavonoids

- Nonribosomal peptides, such as actinomycin-D
- Phenazines
- Phenols
- Polyketide
- Terpenes, including steroids
- Tetrapyrroles.

Research Tools

Enzymes and receptors are often activated or inhibited by endogenous protein, but can be also inhibited by endogenous or exogenous small molecule inhibitors or activators with can bind to the active site or on the allosteric site.

An example is the teratogen and carcinogen phorbol 12-myristate 13-acetate which is a plant terpene which activates protein kinase C promoting cancer, making it a very useful investigative tool. There is also interest in creating small molecule artificial transcription factors to regulate gene expression, examples include wrenchnol (a wrench shaped molecule). Binding of ligand can be characterised using a variety of analytical techniques such as surface plasmon resonance, microscale thermophoresis or dual polarisation interferometry to quantify the reaction affinities and kinetic properties and also any induced conformational change.

Nucleosides and Nucleotides

Nucleosides are molecules formed by attaching a nucleobase to a ribose ring. Examples of these include cytidine, uridine, adenosine, guanosine, thymidine and inosine.

Nucleosides can be phosphorylated by specific kinases in the cell, producing nucleotides. Both DNA and RNA are polymers, consisting of long, linear molecules. The repeating structural units, or monomers, of the nucleic acids are called nucleotides.

Each nucleotide is made of an acyclic nitrogenous base, a pentose and one to three phosphate groups. They contain carbon, nitrogen, oxygen, hydrogen and phosphorus. They serve as sources of chemical energy (adenosine triphosphate and guanosine triphosphate), participate in cellular signaling (cyclic guanosine monophosphate and cyclic adenosine monophosphate), and are incorporated into important cofactors of enzymatic reactions (coenzyme A, flavin adenine dinucleotide, flavin mononucleotide, and nicotinamide adenine dinucleotide phosphate).

Saccharides

Monosaccharides are the simplest form of carbohydrates with only one simple sugar. They essentially contain an aldehyde or ketone group in their structure. The presence of an aldehyde group in a monosaccharide is indicated by the prefix *aldo-*. Similarly, a ketone group is denoted by the prefix *keto-*.

Examples of monosaccharides are the hexoses glucose, fructose, and galactose and pentoses, ribose, and deoxyribose Consumed fructose and glucose have different rates of gastric emptying, are differentially absorbed and have different metabolic fates, providing multiple opportunities for 2 different saccharides to differentially affect food intake. Disaccharides are formed when two monosaccharides, or two single simple sugars, form a bond with removal of water. They can be hydrolyzed to yield their saccharin building blocks by boiling with dilute acid or reacting them with appropriate enzymes.

Examples of disaccharides include sucrose, maltose, and lactose. Polysaccharides are polymerized monosaccharides, complex, carbohydrates. They have multiple simple sugars. Examples are starch, cellulose, and glycogen. They are generally large and often have a complex branched connectivity.

Because of their size, polysaccharides are not water-soluble, but their many hydroxy groups become hydrated individually when exposed to water, and some polysaccharides form thick colloidal dispersions when heated in water. Shorter polysaccharides, with 3 - 10 monomers, are called oligosaccharides. A fluorescent indicator-displacement molecular imprinting sensor was developed for discriminating saccharides. It successfully discriminated three brands of orange juice beverage. The change in fluorescence intensity of the sensing films resulting is directly related to the saccharide concentration.

Lignin

Lignin is a random polymer composed mainly of aromatic rings with short (up to three) aliphatic carbons chains connecting the rings. Lignin is the second most common biopolymer (after cellulose) and is one of the primary structural components of most plants. It contains subunits derived from *p*-coumaryl alcohol, coniferyl alcohol, and sinapyl alcohol and is unusual among biomolecules in that it is racemic i.e. it is not optically active.

The lack of optical activity is because the polymerization of lignin occurs via free radical coupling reactions in which there is no preference for either configuration at a chiral center.

Lipids

Lipids are chiefly fatty acid esters, and are the basic building blocks of biological membranes. Another biological role is energy storage (e.g., triglycerides). Most lipids consist of a polar or hydrophilic head (typically glycerol) and one to three nonpolar or hydrophobic fatty acid tails, and therefore they are amphiphilic. Fatty acids consist of unbranched chains of carbon atoms that are connected by single bonds alone (saturated fatty acids) or by both single and double bonds (unsaturated fatty acids). The chains are usually 14-24 carbon groups long, but it is always an even number. For lipids present in biological membranes, the hydrophilic head is from one of three classes:

- Glycolipids, whose heads contain an oligosaccharide with 1-15 saccharide residues.
- Phospholipids, whose heads contain a positively charged group that is linked to the tail by a negatively charged phosphate group.
- Sterols, whose heads contain a planar steroid ring, for example, cholesterol.

Other lipids include prostaglandins and leukotrienes which are both 20-carbon fatty acyl units synthesized from arachidonic acid. They are also known as fatty acids

Amino Acids

Amino acids contain both amino and carboxylic acid functional groups. (In biochemistry, the term amino acid is used when referring to those amino acids in which the amino and carboxylate functionalities are attached to the same carbon, plus proline which is not actually an amino acid). Amino acids are the building blocks of long polymer chains. With 2-10 amino acids such chains are called peptides, with 10-100 they are often called polypeptides, and longer chains are known as proteins. These protein structures have many structural and enzymatic roles in organisms.

There are twenty amino acids that are encoded by the standard genetic code, but there are more than 500 natural amino acids. When amino acids other than the set of twenty are observed in proteins, this is usually the result of modification after translation (protein synthesis). Only two amino acids other than the standard twenty are known to be incorporated into proteins during translation, in certain organisms:

- Selenocysteine is incorporated into some proteins at a UGA codon, which is normally a stop codon.

- Pyrrolysine is incorporated into some proteins at a UAG codon. For instance, in some methanogens in enzymes that are used to produce methane.

Besides those used in protein synthesis, other biologically important amino acids include carnitine (used in lipid transport within a cell), ornithine, GABA and taurine.

Protein Structure

The particular series of amino acids that form a protein is known as that protein's primary structure. This sequence is determined by the genetic makeup of the individual. Proteins have several, well-classified, elements of local structure formed by intermolecular attraction, this forms the secondary structure of protein.

They are broadly divided in two, alpha helix and beta sheet, also called beta pleated sheets. Alpha helices are formed of coiling of protein due to attraction between amine group of one amino acid with carboxylic acid group of other.

The coil contains about 3.6 amino acids per turn and the alkyl group of amino acid lie outside the plane of coil. Beta pleated sheets are formed by strong continuous hydrogen bond over the length of protein chain. Bonding may be parallel or antiparallel in nature.

Structurally, natural silk is formed of beta pleated sheets. Usually, a protein is formed by action of both these structures in variable ratios. Coiling may also be random. The overall 3D structure of a protein is termed its tertiary structure. It is formed as result of various forces like hydrogen bonding, disulfide bridges, hydrophobic interactions, hydrophilic interactions, van der Waals force etc.

When two or more different polypeptide chains cluster to form a protein, quaternary structure of protein is formed. Quaternary structure is a unique attribute of polymeric and heteromeric proteins like hemoglobin, which consists of two alpha and two beta peptide chains.

Apoenzymes

An apoenzyme is the inactive storage and generally secretory form of a protein. This is required to protect the secretory cell from the activity of that protein.

Apoenzymes becomes active enzyme on addition of a cofactor. Cofactors can be either inorganic (e.g., metal ions and iron-sulfur clusters) or organic compounds, (e.g., flavin and heme). Organic cofactors can be either prosthetic groups, which are tightly bound to an enzyme, or coenzymes, which are released from the enzyme's active site during the reaction.

Isoenzymes

Isoenzymes are enzymes with similar function but different structure. They are products of different genes. They are produced in different organs to perform the same function. LDH are examples of such enzymes. Their varied levels in blood are used to determine any deformity in the organ of secretion.

Vitamins

A vitamin is a compound that is generally not synthesized by a given organism but is nonetheless vital to its survival or health (for example coenzymes). These compounds must be absorbed, or eaten, but typically only in trace quantities. When originally proposed by Casimir Funk, a Polish biochemist, he believed them to all be basic and therefore named them vital amines. The "l" was later dropped to form the word vitamines.

Chapter 3

Plant Breeding

All present forms of life are the product of three factors:

- mutation, the fundamental source of heritable variation,
- environmental factors, which influence the selection of those mutations that survive and reproduce, and
- time, during which the genotype and environment constantly interact and evolutionary change is realized.

Genetic variation found in nature does not represent the original spectrum of spontaneous mutations. Rather, this is the result of genotypes recombining in populations and continuously interacting with environmental forces. Green plants are the ultimate source of resources required for human life, food, clothing, and energy requirements. Prehistoric people, who depended on their skills as hunters, drew upon abundant natural vegetation to collect nutritious and nonpoisonous fruits, seeds, tubers, and other foods.

As human populations increased, greater and safer supplies of food had to be found, and gradually production systems based on plant domestication were developed. The domestication of crops historically has been influenced by ecological and agricultural conditions, as well as by food gathering preferences. Genotypes that have adapted to a wide range of climatic and edaphic conditions typically have been selected for cultivation.

The achievement of higher yielding crops facilitated population growth, sedentary settlements, and further development. Which crops were domesticated depended not only on the number of seeds or the size of fruits, but also on taste, palatability, and other factors.

Only a small fraction of the world's approximately 200 000 plant species have been found suitable for domestication; humans have used about 3000 of these for food, fibre, spices, etc., with 200 ultimately domesticated as crops.

Today, only 15-20 of these are food crops of major importance. The means of developing new plant varieties for cultivation and use by humans has come to be called plant breeding. Early on, it primarily involved selection, the choice between good and bad plants. People learned not to eat all the "best fruit" but to plant the seed from some of them. Genetics became a fundamental science of plant breeding after the Moravian monk J.G. Mendel discovered the laws of heredity in the mid-19th century.

Plant breeding further advanced when the methodology of hybridization was developed. Its aim was to combine various desirable properties of many plants in one plant, instead of just choosing between good and bad plants. This method, often supplemented by germplasm derived from induced mutation, has become the most common one for breeding plants through sexual reproduction.

However, some crops—including bananas, apples, cassava, and sugar cane—reproduce vegetatively, especially those that are fully sterile without seeds. For this important group, alternative approaches had to be developed, namely techniques of manipulation with somatic tissue: mutation breeding and biotechnology.

Mutation Breeding

Plant breeding requires genetic variation of useful traits for crop improvement. Often, however, desired variation is lacking. Mutagenic agents, such as radiation and certain chemicals, then can be used to induce mutations and generate genetic variations from which desired mutants may be selected.

Mutation induction has become a proven way of creating variation within a crop variety. It offers the possibility of inducing desired attributes that either cannot be found in nature or have been lost during evolution. When no gene, or genes, for resistance to a particular disease, or for tolerance to stress, can be found in the available gene pool, plant breeders have no obvious alternative but to attempt mutation induction.

Treatment with mutagens alters genes or breaks chromosomes. Gene mutations occur naturally as errors in deoxyribonucleic acid (DNA) replication. Most of these errors are repaired, but some may pass the next cell division to become established in the plant offspring as spontaneous mutations. Although mutations observed in a particular gene are rare, there are probably 100 000 genes in a cell of a higher plant. This means that every plant may carry one or more spontaneous mutations into the next generation.

Gene mutations without phenotypic (visible) expressions are usually not recognized. Consequently, genetic variation appears rather limited, and scientists have to resort to mutation induction. There are no other economic ways of altering genes, except to wait a long time for spontaneous mutations to occur.

Artificial induction of mutations by ionizing radiation dates back to the beginning of the 20th century. But it took about 30 years to prove that such changes could be used in plant breeding. Initial attempts to induce mutations in plants mostly used X-rays: later, at the dawn of the "Atomic Age", gamma and neutron radiation were employed as these types of ionizing radiations became readily available from newly established nuclear research centres.

Major efforts were devoted during this initial phase of mutation induction to define optimal treatment conditions for reproducibility. Research focused on changing "random" mutation induction into a more directed mutagenesis to obtain more desirable and economically useful mutations. However, it did not lead to the desired alterations in the mutant spectrum. Limitations were the concomitant increase of plant injury with increasing radiation dose and the low frequency of economically useful mutations. This led scientists to search for potentially better mutagens. As a result, new methods of radiation treatment, as well as chemical agents with mutagenic properties, were found.

Plant Biotechnology

Breeding for improved plant cultivars is based on two principles: genetic variation and selection. The process is extremely labourious and time consuming with high inputs of intellectual and manual work. However, the development of plant cell and tissue culture over the last 20 years has made it possible to transfer part of the breeding work from field to laboratoryconditions.

Extensive research has resulted in new areas of plant breeding, namely "plant biotechnology" and "genetic engineering". They are based on cellular totipotency or the ability to regenerate whole, flowering plants from isolated organs (meristems). Pieces of tissue, individual cells and protoplasts. The isolated plant parts are aseptically grown in test tubes on artificial media. For many developing countries, breeding crops for tolerance to salinity and acidity in soils is of high priority. Current breeding strategies (including mutation induction and *in vitro* selection) have clearly been successful in incorporating degrees of tolerance in different species. The use of genetic engineering for creating environmental stress-resistant plants will depend on the identification of specific genes which contribute to the adaptation to specific stress environments.

In tropical countries, agriculture practices have maintained the yield level of different crops through "intercropping" instead of by increased monocrop cultivation. Breeding crops for multiple functions—such as biomass production, improved soil and water practice, and composting —is a desirable support of sustainable agriculture in developing countries. The mixed planting of a main crop with specific cover crops (e.g. forage legumes or grasses) minimizes the use of herbicides.

Role of the Seibersorf Laboratories

The Plant Breeding Unit of the IAEA Laboratories at Seibersdorf was set up in the mid-1960s to support the Joint FAO/IAEA Division's programme of genetic crop improvement. Nuclear techniques in plant breeding are developed and transferred to countries by research and development in mutation breeding and related biotechnological techniques, training scientists from developing countries, and providing irradiation services and technical advice. Initial research in the Plant Breeding Unit focused on the development of mutation induction methods with ionizing radiation and chemical mutagens.

The aim was to achieve high mutagenic efficiency, i.e., a high frequency of desirable mutations at minimal plant injury and the highest possible reproducibility. This required a definition of radiation source characteristics in terms of dose homogeneity and precise assessment of absorbed dose in biological targets by appropriate dosimetry. Irradiation of seeds with gamma rays and neutrons was commonly done, given the ease of handling, the simple standardization of factors which modify radiation sensitivity, and good reproducibility.

The establishment of methods for controlling oxygen-dependent effects in the radiobiological response to electromagnetic radiation was a major achievement. The Laboratory actively contributed to standardizing neutron irradiation of seeds in nuclear reactors by developing special facilities for this purpose. These were known as SNIP, for Standard Neutron Irradiation Facility for swimming-pool-type reactors; and as USIF, for Uranium Shielded Irradiation Facility for Triga-type reactors. This research was the basis for the IAEA Laboratories' worldwide seed irradiation service using fast and thermal neutrons at a high-dose precision and reproducibility of induced effects.

Moreover, efficient and accurate treatments of seeds with chemical mutagens, mostly alkylating agents and azides, were developed with the aid of isotope-labelled compounds and compared with mutation induction by ionizing radiation. The Unit has undertaken supportive research on mutation breeding in cereals, pulse crops, industrial crops, and vegetatively propagated crops.

As each crop species has a variable reproductive capacity (number of progenies per plant) and a specific system of reproduction (self- or crosspollinated sexual reproduction or asexual propagation), a universal breeding approach cannot be developed and species-specific procedures have to be applied.

Most vegetatively or asexually propagated species are difficult to improve genetically by conventional cross- and mutation breeding methods. These breeding problems can be more easily resolved by using biotechnology in combination with mutation induction, and the Unit initiated *in vitro* mutation breeding activities during the mid-1980s. Several tropical food crops of great importance to the food security of developing countries were chosen as the main focus of R&D and training activities in biotechnological plant breeding at the IAEA Laboratories.

Research and Development Activities

The Unit provides focused support to the FAO/IAEA's co-ordinated research and technical co-operation programmes. Assistance is provided to numerous projects in terms of expertise for building facilities for plant tissue culture and mutagenic treatment, for quality control of dosimetry of mutagenic irradiation, and for the development and transfer of nuclear technologies for plant improvement.

Ongoing R&D includes the application of nuclear methods and associated advanced techniques, such as *in vitro* culture and molecular genetics, to improve the production of a wide range of crops through mutation breeding. The development of biotechnological methods for breeding vegetatively propagated crop plants of major importance in developing countries has a high priority.

Currently, the following R&D areas are being pursued:

Somaclonal and mutagen induced variation: Systematic studies are being conducted to compare the genetic variation caused by tissue culture (somaclonal) variation with that induced by irradiation and chemical agents. Genetic variation is being studied among maize plants derived from *in vitro* cultured material via somatic embryogenesis. This is being done to assess the nature of somaclonal and induced variation and its potential for use in practical breeding.

Mutation induction and breeding technology for banana and plantain: Low genetic variation and sterility handicap genetic improvement of banana and plantains *(Musa spp.)* by conventional breeding techniques. Shoot-tip culture and *in vitro* plant regeneration are being investigated for use in mutation induction and mutant selection.

Somatic embryogenesis and plant regeneration from cell suspensions of *Musa* are used to develop somatic cell manipulation procedures for banana and plantain breeding.

Methods of screening such plants for resistance to Panama disease are studied in tissue culture, and biochemical markers (peroxidase) are applied for the identification of tolerant genotypes. DNA markers are used for identifying mutants and characterizing cultivars and species *of Musa*. Mutant clones identified at the Seibersdorf Laboratories are tested in the field in tropical countries.

Mutation breeding to improve the tolerance to environmental stress of Azolla: *Azolla* is a small aquatic fern that lives in symbiotic relationship with the nitrogen-fixing cyanobacterium *Anabaena*. Under suitable field conditions *Azolla* can double its weight every 3-5 days. The *Azolla-Anabaena* symbiotic system provides green manure for flooded crops, particularly rice. Induced mutagenesis has produced *Azolla* variants tolerant to high salinity, toxic aluminium levels, and/or to herbicides.

Tolerant plants are being investigated under field conditions to confirm that heritable changes cause the increased tolerance to environmental stress.

Methods of mutation induction and breeding of tropical root and tuber crops (cassava and yam): Cassava and yam are among the most important staple food crops of the lowland tropics. Mutation breeding technology is being developed to increase variation in plant stature, cyanide content, disease, and pest resistance.

In vitro techniques are used for the propagation of healthy plants and improved clones. Somatic embryogenesis is being developed for cassava and yam improvement through *in vitro* mutagenesis and later on by somatic cell manipulation. Mutant and polyploid clones are prepared for field testing in Member States.

Tissue culture in cocoa as a system for more efficient mutation breeding: Attempts to breed cocoa for disease resistance have yielded very limited success. A major constraint is that little variation exists in currently available cultivars. Somatic embryogenesis is being developed for propagation of desirable genotypes and, through *in vitro* mutagenesis and pollen mutagenesis, is being applied for induction of virus-resistant cocoa trees in Ghana. Plant breeding research at the Seibersdorf Laboratories is directly problem- and clientoriented.

Many positive results of scientific work have been achieved by junior scientists from developing countries during their assignments under the IAEA's fellowship training programme. Local cultivars and genetic material from tropical countries are brought to the Seibersdorf Laboratories, transferred to tissue culture conditions and used for experimental work. Protocols and techniques that are specifically developed for a crop and a particular genotype are then directly used in national programmes. Additionally, breeding material originating from mutant lines and clones which are ready for field testing are dispatched from Seibersdorf to developing Member States in support of their breeding programmes.

Training of Plant Breeders

Training in plant breeding represents the most active component of technology transfer at the Seibersdorf Laboratories. For 20 years the Plant Breeding Unit has supported the Agency's fellowship programme and organized interregional training courses. Training activities are closely connected with R&D efforts on crop plant improvement and the application of nuclear techniques in breeding.

During a period of three to twelve months, fellows usually work with radiation or chemical induced mutagenesis in plant species cultivated in their home countries. Whenever possible, training of small groups of two-to-five fellows is organized for solving common problems. The experiments are individually designed to assure that laboratory techniques and results will be directly applicable upon return to the home institute.

As a result of their work, fellows have produced numerous scientific publications in internationally recognized journals and symposia proceedings. Very often, as continuation of a fellowship in Seibersdorf, fellows participate in co-ordinated research and technical co-operation projects of the IAEA.

The FAO/IAEA Interregional Training Course on "Induction and Use of Mutations in Plant Breeding" has been held at the Seibersdorf

Laboratories since 1982. Twenty participants from different Member States of FAO and IAEA are admitted annually to this intensive training course that usually lasts 6 to 8 weeks.

Through lectures, laboratory exercises, field experiment evaluations, seminars, and excursions, participants are made aware of the latest advanced mutation techniques and biotechnological and molecular biology methods for crop improvement. Special training is given in the safe handling of radiation sources, radioisotopes, and particularly hazardous mutagenic chemicals. At the end of each course, participants are able to discuss and evaluate the potential role of induced mutations and advanced biotechnologies in their national breeding programmes for specific crop improvement of cereals, legumes, oil crops, forages, vegetables, fruits, root and tuber crops, palms, rubber, and other plants.

Support for National Programmes

A radiation treatment service is provided at no cost to FAO and IAEA Member States to foster the application of nuclear techniques in crop improvement programmes and to render direct support to plant breeders in developing countries. Mutagenic treatment is applied to seeds, corms, tubers, scions, cuttings, and tissue cultures *("in vitro* materials") with precise doses of gamma and fast neutron radiation. The doses are carefully calibrated to assure reproducible effects. Users of the service are requested to report on the objectives of the applied mutation breeding project and to provide an adequate material (population size) to ensure a high probability for mutation induction of desired characters.

Moreover, a prior radiosensitivity test in a greenhouse is frequently performed to assess useful radiation doses for the great variety of biological samples in mutation breeding. The treated materials are dispatched with a detailed irradiation protocol and with the request to report on the induced radiation effects in the first and second mutation generation. This feedback is required to improve radiosensitivity estimates of species and cultivars from different environments.

Over the last 25 years, the Unit has provided radiation services on more than 20 000 samples from the majority of Member States from the FAO and IAEA. Most of these were seed samples which were irradiated with cobalt-60 gamma rays. Recently, however, requests for mutagen treatment of *in vitro* materials and for fast neutrons have become more frequent. This reflects the increasing importance of biotechnology and molecular genetics in plant improvement programmes.

Less than 80 mutant varieties were officially released before the start of irradiation services. Over the past quarter century, more than 1500 cultivars of crop plants and ornamentals with significantly improved attributes — increased yield, improved quality, higher market value, disease resistance, and/or stress tolerance —have been released. Some of these mutant varieties were derived from radiation services provided by the Seibersdorf Laboratory.

Cell Suspension Culture

There are two basic systems for growing cells in culture, as monolayers on an artificial substrate (i.e., adherent culture) or free-floating in the culture medium (suspension culture).

Characteristics of Each Culture Type

The majority of the cells derived from vertebrates, with the exception of hematopoietic cell lines and a few others, are anchorage-dependent and have to be cultured on a suitable substrate that is specifically treated to allow cell adhesion and spreading (i.e., tissue-culture treated). However, many cell lines can also be adapted for suspension culture. Similarly, most of the commercially available insect cell lines grow well in monolayer or suspension culture.

Cells that are cultured in suspension can be maintained in culture flasks that are not tissue-culture treated, but as the culture volume to surface area is increased beyond which adequate gas exchange is hindered (usually 0.2 – 0.5 mL/cm_2), the medium requires agitation. This agitation is usually achieved with a magnetic stirrer or rotating spinner flasks.

Meristem Culture

Culture of meristem tips were started first by Morel and co-workers in the nineteen fifties with the intention of obatining virus -free plants from those already infected. In 1960 he tried this methods in a virus infected Cymbidiuma and succeeded in ontaining a aplantling free of infection. Following this, Wimber (1963),Hamilton (1964) and Morel (1964) developed methods by which a single detached meristem could be cultured, divided into several bits and each bit subcultured by repeating this several times over, an incredibly large number of plantings could be obtained from a single protocorn in one year.

Meristems are dissected out with the help of a sharp razor-blade, the operation being carried out under a microscope. It should be 500 in length and about the same in diameter. These tiny bits of meristems are then sterilised with 2 % calcium hypochlorite solution and placed in a liquid nutrient medium in a flask under sterile conditions. The flask is then placed on a rotary shaker and rotated at a given speed. This is to prevent formation of any polarity to the developing meristem. Light equivalent to 200 foot candles and a temperature of 22 C are maintained throughout.

Within a week the meristems will turn green. It starts developing several lumbs of tissue. At this stage the protocorn is taken out, washed and cut up into as many pieces as there are balls of tissue. Each bit is dipped in a sterilant and put in a tube containing fresh liquid medium and agitated. The procedure is repeated until as many plantings as required are produced. The tiny balls of tissue are then transferred to normal nutrient agar medium and akept static as in seedling culture. Each ball of tissue, after reaching 4 mm in length differentiates into an individual plant.

The technique of propagation by meristem culture provided a great boost to the orchid industry. Rare plants of special qualities which could not be propagated by seeds due to genetic factors, could be multiplied easily and that too at a fantastic rate. This naturally brought down the prices of these orchids which were otherwise too expensive for most people to possess. Precious plants which fell victims to the incurable virus diseases also could be salvaged.

The only disadvantage with propagation by meristem culture is that removal of meristem causes temporary set-back in the growth of the mother plant. To avoid this,parts of seeds due to genetic factors, could be multiplied easily and that too at a fantastic rate. This naturally brought down the prices of these orchids which were otherwise too expensive for most people to possess. Precious plants which fell victims to the incurable virus diseases also could be salvaged.

The only disadvantage with propagation by meristem culture is that removal of meristem causes temporary set-back in the growth of the mother plant. To avoid this, parts of the plant other than the apical meristem, have been tried in tissue culture experiments.Churchill, Arditti and Ball (1971) conducted experiments with leaf-tips and reported that the excised leaf-tips of Epidendrum and Laeliocattleya produced calli when put in liquid culture medium and roated. Upon being transferred to agar media these calli produced plantings. One of the more recent development in vegetative multiplication of orchids is the use of cytokinis for promoting dormant buds to develop into young plants or 'Keikis. This has been well described in an article by Brasch and kocsis (1980) from the McMaster University, Hamilton, Ontario, Canada. Experiments are in progress on several genera, but the greatest success so far has been achieved in Phalaenopsis, which has a mopodial habit. While the mericloning technique is wiedly used by large establishmets like commercial nurseries, for the amateur growing a few plants and desiring to increase them, the discovery of "Keiki grow", and its availability are of great interest. Orchid "keiki Fix " formulated by N.F.Hass and marketted by H.G. Hees, 99 a kilin Ride, workingham, Berks, RG11 3 PD, U. K is reported to be equally effective on the dormant buds on Phalaenopsis inflorescence stalk.

Diseases and Pests

Likey other group of cultivated plants, orchids also are prone to a number of diseases caused by various agents like viruses, fungi, bacteria, insect pests etc. All bacterial, fungal and virus diseases are infectious and unless checked in time, the disease will spread from plant to plant. Except for the viruses, effective preventives and remedies have been discovered for these ailments and if detected in time, they could successfully be eliminated before serious damage is done to the plant.

Preventive measures should be followed strictly and according to rules. Most fungi and bacteria thrive under damp and soggy conditions, because invariably the spores need moisture for germination. Hence overcrowding should be avoided in orchid houses and facilities provided for free air circulation. Debris should not be allowed to accumulate in the house or in its near vicinity. Watering, spraying, fogging etc. should be done early in the day so that by night,when usually the temperature comes down, teh plants may remain completely dry and relatively impervious to attack by fungi or bacteria.

The grower has to be always on the alert for the symptoms of disease in plants. They should be examined daily so that even the slightest symptom is not overlooked and immediate steps are taken in order to check the spread of the disease.

Virus Diseases

Viruses and the diseases they produce are the most difficult to tackle, since our knowledge about them is far from complete, and whatever little information we have is often confusing. This is because viruses are not specific or reliable in the reactions they produce in different species of plants, e.g.. Cymbidium Mosaic Virus produces mosaic mottle on leaves of Cymbidium, whereas in Cattleya it produces necrotic spots (dead spots), which is entirely different in appearance from the mosaic mottle. The reverse situation may also occur in infected plants, i.e. two different viruses producing two different diseases, may induce the same lesions when they occur in the same plant. Another peculiarity of the virus disease is that, when the infection is mild, the virus may remain inside the plant for years together, without any visible symptoms, but always to be reckoned with as a dangerous sources of fresh infections. At other times the infection may be so virulent that the plants are killed within a few weeks.

Altogether about 32 virus diseases are known to occur in orchids (Jensen, 1959). But this does not necessarily mean that orchids are susceptible to thirty two different kinds of viruses. The same virus may produce more than one disease in different species of orchids. Among orchid viruses, Cymbidium Mosaic Virus (CyMV), Odontoglossum Ring Spot Virus (ORsV) and Dendrobium nobile Mosaic Virus are reported to infect non-orchidaceous plants as well.

CyMV, Tobacco Mosaic Virus-O (TMV-O) and ORsV are the most common among orchid viruses. Most of the cultivated genera are susceptible to these three viruses. CyMV is a flexuous rod 18m in diameter and 500 m in length as shown by Gold (1950), Gold and Jensen (1951) and Jensen and Gold (1955). TMV-O is related to,but not the same as, the Tobacco Mosaic Virus which infects also the potato plant. It is a straight short road about 300 m in length. According to Kado et al. (1968) it is serologically the same as the TMV has a wide range of hosts whereas TMV-O infects only a limited number of plants. ORsV was identified by Paul and Wetter (1965) as a rigid short rod. According to Kado, this virus is the same as TMV-O (Northen, 1970).

Callus Culture

The callus culture is a technique of tissue culture, it is usually carried out on solidified gel medium in the presence of growth regulators and initiated by inoculation of small explants or sections from established organ or other cultures (the inocula).

Callus Culture and Regeneration

Regeneration of plants by micropropagation of h vitro cultures can be achieved from organ primordia existing in shoot tips and axillary bud explants. Alternatively, plants can be regenerated from unorganized callus tissues derived from different explants by dedifferentiation induced by exogenous growth regulators. Plant regeneration from calli is possible by *de novo* organogenesis or somatic embryogenesis. Callus cultures also facilitate the amplification of limiting plant material. In addition, plant regeneration from calli permits the isolation of rare somaclonal variants which result either from an existing genetic variability in somatic cells or from the induction of mutations, chromosome aberrations, and epigenetic changes by the in vitro applied environmental stimuli, including growth factors added to the cultured cells.

In *Arabidopsis thaliana,* one of the earliest studies on callus formation was conducted by Loewenberg, who grew seedlings in a medium containing kinetin and parachlorophenoxyacetic acid. Subsequent studies of cell culture and regeneration demonstrated that it is relatively easy to regenerate *Arabidopsis* from callus cultures by many approaches.

However, the efficiency of regeneration and the proportion of somaclonal variants are considerably influenced by the ecotype and the source of explants, as well as by the medium and growth regulators employed. Hypocotyl and root explants thus provide excellent materials for callus initiation and regeneration in contrast to stem and leaf explants.

The tissue culture protocols include callus induction in auxin containing media. Shoot regeneration is induced by lowering the auxin content and increasing the cytokinin levels in the media. The elongated shoots are usually transferred into root induction media, but even in the absence of roots, flowering and seed setting can take place in test tubes in vitro. The method described here utilizes root explants that can be cultured and efficiently regenerated in large quantities. The described technique can be applied equally well for the Arabidopsis ecotypes Columbia, C24, RLD, and Wassilewskija. Without changing the general protocol, some slight alterations may be required for other ecotypes or mutants of Arabidopsis.

Materials

1. Dry seeds of Arabidopsis thaliana.
2. 10% Sodium hypochlorite solution containing 0.1% Triton X-100.
3. Eppendorf tubes.
4. Sterilized double-distilled water.
5. Culture tubes.
6. 9 cm Petri dishes.
7. Pair of forceps.
8. Scissors or scalpel blades.

9. Sterile 250-mL Erlenmeyer flasks with plugs.
10. Rotary shaker set at 120 rpm.
11. Microfuge.
12. Basal medium (BM): Murashige and Skoog (MS) medium (6) or MS medium for Arabidopsis (MSAR) medium (7) with 3% sucrose (pH 5.8) can equivalently be used.
13. 0.5X BM: BM consisting of half concentration of MS macroelements and 0.5% sucrose.
14. MSAR I (Callus medium): Supplement BM with 0.5 mg/L 2,4—dichlorc~ phenoxyacetic acid (2,4-D), 2.0 mg/L indole3-acetic acid (IAA), 0.5 mgL 6-(y,y-dimethylallylamino~purine riboside (IPAR), and gel using either 0.8% agar or 0.2% gelrite.
15. MSAR I1 (Shoot medium): Supplement BM with 2.0 mg/L 6-(y,y-dimethylallylamino)- purine riboside (IPAR), 0.05 mg/L a-naphtaleneacetic acid (NAA), and either 0.8% agar or 0.2% gelrite.
16. MSAR 111 (root inducing medium): Supplement BM with 1.0 mg/L IAA, 0.2 mg/L indole3-butyric acid (IBA), 0.2 mg/L 6-furfuryl-aminopurine (kinetin), and either 0.8% agar or 0.2% gelrite.

Methods

1. Surface sterilize 0.1 g (approx 5000) seeds in Eppendorf tubes by adding a 1 mL of 10% (v/v) solution of sodium hypochlorite containing 0.1% Triton X-100 as surfactant and shaking for about 15 min.
2. Pellet the seeds by slow centrifugation in a microfuge for a few seconds and remove the supernatant. Wash the seeds five times with 1 mL of sterile water.
3. Germinate the seeds in Petri dishes containing 0.8% agar gelled 0.5X BM medium (BM medium containing half concentration of macroelements) using 16 h light and 8 h dark period at 25°C.
4. Place 15-20 1-wk-old seedlings into 250-mL Erlenmeyer flasks containing approx 35 mL of liquid BM medium with 3% sucrose. 5. Place the flasks on a rotary shaker at 120 rpm using 16 h light and 8 h dark period at 25°C.
6. Harvest the roots of these seedlings after 15-20 d of growth.
7. Place the roots into a Petri dish, remove all liquid and cut the roots into small pieces.

8. Transfer root explants onto MSAR I plates and either cover them with aluminium foil or place them in low light conditions.
9. After 3 wk, transfer the root-derived calli either to fresh MSAR I plates for maintenance or to MSAR I1 plates for shoot regeneration.
10. After 2 wk, pick the regenerating shoots from the calli and transfer them into glass jars with MSAR I1 medium for further elongational growth.
11. Shoots consisting of 4-6 leaves may be transferred further into MSAR 111 medium for 3-6 d to induce root development.
12. Transfer the elongated shoots (approx 2 cm) into culture tubes containing 0.5 BM agar medium with 0.5% sucrose for rooting, flowering, and setting seeds.

Notes

1. Stocks of growth regulators are prepared at 1 mg/mL concentration, filter sterilized, and stored at 4OC. Growth regulators are added to autoclaved medium after it has cooled to about 60°C.
2. Do not sterilize too many seeds in one Eppendorf tube, as they may not be easily dried. In case seeds clump, use a sterile toothpick to break the clumps before plating the seeds.
3. Calli can also be obtained from germinating seeds that are placed on MSAR I plates after surface sterilization.
4. After 15-20 d of culture, 3-5 g (fresh weight) of roots should be produced in each flask. The roots should be white, actively growing, and not yellow-brown or green.
5. At this stage, it is possible to regenerate and amplify shoots from the regenerated leaves by cutting them into pieces and placing them on MSAR I1 Medium.
6. The culture tubes are capped with loose cotton to facilitate the aeration required for seed setting and maturation. Take care not to place the tubes too close to the light source because this will cause moisture condensation inside the tubes and result in a low rate of fertilization.
7. Rooted plants can be transferred to soil after washing the roots with water containing a fungicide, such as Benomyl (0.02%). The plants are gradually acclimated by reducing the humidity of transfer chambers stepwise.

Somaclonal Variation

Somaclonal variation It is the term used to describe the variation seen in plants that have been produced by plant tissue culture. Chromosomal rearrangements are an important source of this variation. Somaclonal variation is not restricted to, but is particularly common in plants regenerated from callus. The variations can be genotypic or phenotypic, which in the latter case can be either genetic or epigenetic in origin. Typical genetic alterations are: changes in chromosome numbers (polyploidy and aneuploidy), chromosome structure (translocations, deletions, insertions and duplications) and DNA sequence (base mutations). Typical epigenetic related events are: gene amplification and gene methylation.

If no visual, morphogenic changes are apparent, other plant screening procedures must be applied. There are both benefits and disadvantages to somaclonal variation. The phenomenon of high variability in individuals from plant cell cultures or adventitious shoots has been named somaclonal variation.

Benefits

The major likely benefit of somaclonal variation in plant is improvement. Somaclonal variation leads to the creation of additional genetic variability. Characteristics for which somaclonal mutants can be enriched during *in vitro* culture includes resistance to disease pathotoxins, herbicides and tolerance to environmental or chemical stress, as well as for increased production of secondary metabolites. Micropropagation can be carried out throughout the year independent of the seasons.

Disadvantages

A serious disadvantage of somaclonal variation occurs in operations which require clonal uniformity, as in the horticulture and forestry industries where tissue culture is employed for rapid propagation of elite genotypes. Ways of reducing somaclonal variation: Different steps can be used. It is well known that increasing numbers of subculture increases the likelihood of somaclonal variation, so the number of subcultures in micropropagation protocols should be kept to a minimum. Regular reinitiation of clones from new explants might reduce variability over time. Another way of reducing somaclonal variation is to avoid 2,4-D in the culture medium, as this hormone is known to introduce variation. Vitrification [hyperhydracity] may be a problem in some species. In case of forest trees, mature elite trees can be identified and rapidly cloned by this technique. High production cost has limited the application of this technique to more valuable ornamental crops and some fruit trees.

Plant Embryogenesis

Plant embryogenesis is the process that produces a plant embryo from a fertilised ovule by asymmetric cell division and the differentiation of undifferentiated cells into tissues and organs. It occurs during seed development, when the single-celled zygote undergoes a programmed pattern of cell division resulting in a mature embryo. A similar process continues during the plant's life within the meristems of the stems and roots.

Seeds

Embryogenesis occurs naturally as a result of sexual fertilization and the formation of the *zygotic embryos.* The embryo along with other cells from the motherplant develops into the seed or the next generation, which, after germination, grows into a new plant. Embryogenesis may be divided up into two phases, the first involves morphogenetic events which form the basic cellular pattern for the development of the shoot-root body and the primary tissue layers; it also programs the regions of meristematic tissue formation.

The second phase, or postembryonic development, involves the maturation of cells, which involves cell growth and the storage of macromolecules (such as oils, starches and proteins) required as a 'food and energy supply' during germination and seedling growth. Embryogenesis involves cell growth and division, cell differentiation and programed cellular death. The zygotic embryo is formed following double fertilisation of the ovule, giving rise to two distinct structures: the plant embryo and the endosperm which together go on to develop into a seed. Seeds may also develop without fertilization, which is referred to as apomixis. Plant cells can also be induced to form embryos in plant tissue culture; such embryos are called *somatic embryos.*

Following fertilization, the zygote undergoes an asymmetrical cell division that gives rise to a small apical cell, which becomes the embryo and a large basal cell (called the suspensor), which functions to provide nutrients from the endosperm to the growing embryo. From the eight cell stage (octant) onwards, the zygotic embryo shows clear embryo patterning, which forms the main axis of polarity, and the linear formation of future structures. These structures include the shoot meristem, cotyledons, hypocotyl, and the root and root meristem: they arise from specific groups of cells as the young embryo divides and their formation has been shown to be position-dependent. In the globular stage, the embryo develops radial patterning through a series of cell divisions, with the outer layer of cells differentiating into the 'protoderm.' The globular embryo can be thought of as two layers of inner cells with distinct developmental fates; the apical layer will go on to produce cotyledons and shoot meristem, while the lower layer produces the hypocotyl and root meristem. Bilateral symmetry is apparent from the heart stage; provascular cells will also differentiate at this stage. In the subsequent torpedo and cotyledonary stages of embryogenesis, the embryo completes its growth by elongating and enlarging.

In a dicot embryo, the hypophysis, which is the uppermost cell of the suspensor, differentiates to form part of the root cap. Plant cells can also be induced to form embryos in plant tissue culture; these embryos are called *somatic embryos,* which are used to generate new plants from single cells.

Plant growth and Buds

Embryonic tissue is made up of actively growing cells and the term is normally used to describe the early formation of tissue in the first stages of growth. It can refer to different stages of the sporophyte and gametophyte plant; including the growth of embryos in seedlings, and to meristematic tissues, which are in a persistently embryonic state, to the growth of new buds on stems.

In both gymnosperms and angiosperms, the young plant contained in the seed, begins as a developing egg-cell formed after fertilization (sometimes without fertilization in a process called apomixis) and becomes a plant embryo. This embryonic condition also occurs in the buds that form on stems. The buds have tissue that has differentiated but not grown into complete structures. They can be in a resting state, lying dormant over winter or when conditions are dry, and then commence growth when conditions become suitable. Before they start growing into stem, leaves, or flowers, the buds are said to be in an embryonic state.

Somatic Embryogenesis

Somatic embryos are formed from plant cells that are not normally involved in the development of embryos, i.e. ordinary plant tissue. No endosperm or seed coat is formed around a somatic embryo. Applications of this process include: clonal propagation of genetically uniform plant material; elimination of viruses; provision of source tissue for genetic transformation; generation of whole plants from single cells called protoplasts; development of synthetic seed technology. Cells derived from competent source tissue are cultured to form an undifferentiated mass of cells called a callus.

Plant growth regulators in the tissue culture medium can be manipulated to induce callus formation and subsequently changed to induce embryos to form from the callus. The ratio of different plant growth regulators required to induce callus or embryo formation varies with the type of plant. Asymmetrical cell division also seems to be important in the development of somatic embryos, and while failure to form the suspensor cell is lethal to zygotic embryos, it is not lethal for somatic embryos.

Somaclonal Variation for Disease Resistance

Most Brazilian wheat cultivars have a high yield potential but are susceptible to several diseases such as spot blotch, caused by *Bipolaris sorokiniana*; blast, caused by *Magnaporthe grisea*, and bacterial leaf stripe, caused by *Xanthomonas campestris* pv. undulosa (Xcu). It is believed that resistance to spot blotch and bacterial stripe is governed by polygenes. Although sources of resistance are continually being identified, the degree of resistance of commercial varieties is inadequate. Breeding efforts to transfer resistance from such sources have received top priority during the past two decades, nonetheless, results are discouraging, partly because of the polygenic nature of resistance and partly because of a lack of ample genetic diversity among resistance sources.

The production of wheat haploids via chromosome elimination is one of the latest and most useful techniques in gene transfer experiments. Production of haploid wheat plants and subsequent production of dihaploids (double haploids) can fix desirable characters in a single generation. Utilizing the somaclonal variation technique, attempts have been made to obtain variation in resistance to *B. sorokiniana*, *M. grisea* and Xcu, without drastically altering positive agronomic characteristics of the original genotypes. Our study was aimed at fixing resistance characters of the somaclones by producing haploids and dihaploids through wheat and maize hybridization.

Regeneration of Somaclonal Plants

Success in callus induction and regeneration of somaclonal plants that are resistant to different diseases is obtained by utilizing explants of genotypes that already possess some degree of resistance. The wheat cultivars 'Batuira', 'IAPAR 53', 'IAPAR 40', 'OCEPAR 14', 'OCEPAR 17', 'OCEPAR 19' and 'BR 18', are some of the most popular cultivars in the State of Paraná, Brazil. Since these cultivars were reported to have some degree of resistance to one or the other aforementioned pathogens (Mehta, 1981, 1996a; Mehta *et al.*, 1996), they were selected for somaclonal variation. Seeds obtained from the IAPAR germplasm collection were selected from a typical spike of each wheat cultivar, multiplied by selfing once, then stored for further use.

Fifteen to 20 plants of each cultivar were planted in a glasshouse on three sowing dates separated by a 7-day interval. The floral primordia of these plants in their very initial developmental stages (inflorescences 0.5-2.0 cm in length) were selected for callus induction and regeneration of somaclonal plants, using Petri plates containing 10 ml MS (Murashige and Skoog), callus induction medium, with 2.0 mg 2,4-D/l.

Immature inflorescences were surface sterilized in 40% sodium hypochlorite solution, excised into small pieces of up to 2-3 mm, placed on Petri plates containing callus induction medium and incubated in the dark for 30 days at 26°C. After 30 days, the calli were transferred onto fresh medium for 15 days under similar incubation conditions.

After 45 days, the calli were transferred to a regeneration MS medium (without 2,4-D, but supplemented with 27.8 mg/l $FeSO_4.4H_2O$, 37.3 mg/l Na_2-EDTA, 0.5 mg/l ANA and 3.0 mg/l kinetin) and incubated at 26°C in an 8/16-h dark/light cycle (Araujo *et al.*, 1995). After reaching the three-leaf stage, the regenerated seedlings were climatized by flooding the plates with water and incubated at room temperature for 24 h, followed by transfer to a growth chamber in earthen pots, 20 cm in diameter, containing soil and organic manure. After 30 days pots were fertilized with 4:30:10 N, P_2O_5 and K (2.5 g/pot), and the plants grown to maturity.

Inoculation and Evaluation of Somaclonal Plants for Disease Resistance

Cultures of *B. sorokiniana*, *M. grisea* and Xcu were grown on potato dextrose, oat-meal and nutrient agar, respectively. Cultures were started one week before inoculation for fungal pathogens and 48 h before inoculation for the bacterial pathogen, on Petri plates containing the respective media. After growth, plates were flooded with distilled water and gently scraped; the resulting suspension was filtered through two layers of sterile cheese cloth. The conidial concentration was adjusted to 3.0 x 10^4 conidia/ml and bacterial suspension adjusted to 10^8 cfu (colony-forming units)/ml, by diluting with water. One drop of Tween

20 per 200 ml suspension was added and the suspension was sprayed onto the plants using a small atomizer and pressure pump. Inoculations were started on two leaves of 30-day old plants and each pathogen was inoculated at 10-day intervals using separate leaves of each plant.

The leaves were first washed with distilled water, gently rubbed between the fingers, and each one held upright and given a quick single round of spray, from top to bottom and back again. Inoculations of R2 plants (regenerated second generation somaclones) were made using:

1) a mixture of seven aggressive monosporic isolates of *B. sorokiniana* obtained from the Instituto Agronômico do Paraná (IAPAR) culture collection (strain Nos.: 9130-9134, 9140 and 9141);
2) a mixture of six aggressive isolates of Xcu obtained from the culture collection (strain Nos. 6668, 6712, 6713, 6867, 9217, 9248) one aggressive isolate of *M. grisea* (strain No. 530) from wheat.

The inoculated plants were incubated for 16 h in a completely dark, moist chamber with a water-saturated atmosphere at about 20°C. Later, these were transferred to another growth chamber set at 20°C with about 70% relative humidity and a 12/12-h dark/light period cycle. Wheat cultivar 'Mitacore' plants, highly susceptible to all three pathogens, were also inoculated and maintained as controls.

Leaf area percentage infected (LAI) was calculated 7 days after inoculation by visual estimation using a disease appraisal scale proposed for *B. sorokiniana* and Xcu, where LAI <1% was considered resistant. For *M. grisea*, the 0-1 scale proposed by Zadoks (1972) was used. Disease rating 0 was considered resistant (non-compatible) and 0.25-1.0 as susceptible (compatible) (Prabhu *et al.*, 1992). Resistant plants were grown to maturity.

Production of Haploid Wheat Plants from Wheat X Maize Crosses

Attempts were made to fix resistance characteristics of the somaclonal wheat plants of the R3 generation by producing haploid plants from wheat x maize crosses, following a modified procedure described by Laurie and Bennett (1988). Somaclonal wheat plants were grown in earthen pots in a growth chamber at 20°C. Spikes of the leading tiller were emasculated 2-3 days before anthesis and covered with glycine bags to avoid cross pollination. After 3-4 days, crosses were made using a mixture of freshly collected pollen from glasshouse-grown maize (*Zea mays* L.) hybrids 'XL 321' and 'XL 370', produced by BRASKALB, RS, Brazil. These lines are efficient in producing wheat x maize hybrids (Fernandes, M.I., personal communication). For this purpose, upper and basal spikelets were removed as well as all but primary and secondary florets of the remaining spikelets.

Maize pollen was dusted on the wheat stigmas using a soft paint brush. Pollinations were then repeated the next day. After pollination, spikes were twice injected with 1 ml 0.001% 2,4-D at an interval of 24 h, 4-5 cm below the peduncular node. Caryopses were collected from the spikes 13-14 days post-pollination and surfaces sterilized in a 50% sodium hypochlorite solution for 15 s, followed by rinsing in 75% ethanol for 1 min. Caryopses were excised aseptically under a stereomicroscope and the rescued embryos transferred to test tubes (with scutella facing upwards) containing regeneration medium without 2,4-D. The embryos were incubated in the dark for 4 days at 5°C to break dormancy, and later for 12 days in a growth chamber set at 20°C. Regenerated plantlets (1-2 leaf stage with good root development) were transferred to pots containing vermiculite and kept covered with a glass beaker in a phytotron at 16°C, with 12/12-h light/dark periods. After 7 days the glass beaker was removed, plants were irrigated with nutrient solution and allowed to grow for a few weeks.

After the haploid plants reached the 4-5 tiller stage, their roots were treated for three hours with a solution containing 0.25% colchicine, then rinsed with distilled water. Colchicine-treated plants of the resistant somaclones were then grown to maturity under the conditions previously mentioned for haploid plant production.

Along with the R3 generation somaclones, four elite wheat cultivars 'OCEPAR 16', 'OCEPAR 21', 'OCEPAR 22' and 'IAPAR 60' were also tested for haploid plant production in order to determine whether production of haploids is wheat-cultivar dependent. Simultaneously, the effects of two concentrations of 2,4-D (0.001 and 0.005%) on embryo formation was tested using 50 plants per concentration of each of the two arbitrarily selected wheat cultivars 'Batuira' and 'IAPAR 60'.

Cytological Analysis of Somaclones and their Haploids

Root tips of young seedlings were fixed in 3:1 ethanol acetic acid. After 48 h in fixative, they were Feulgen-stained and examined by light microscopy to determine chromosome numbers. Chromosome counts were made on 15-20 original hexaploid somaclones and 15-20 haploid wheat plants originated from wheat x maize hybrids.

Results and Discussion

Callus initiation was observed within 12-15 days and callus induction frequency varied from about 70-100% depending upon the cultivar. Floral primordia in their advanced stages (>3 cm) did not produce calli. Embryogenic calli were compact and yellowish-green. Although no albino or sterile plants were regenerated, abnormal pseudofoliar structures and excessive root development without shoots were observed for some cultivars. Calli were transferred successively to MS regeneration medium without 2,4-D every 30 days.

Regeneration frequency of R1 varied and was cultivar dependent. Resistance against *B. sorokiniana* and *M. grisea* was observed in some R2 plants, which means that novel resistance genes could be introduced through embryogenesis, which would otherwise be unavailable through conventional breeding methods (De Buyser *et al.*, 1987). None of the somaclonal plants were resistant to Xcu.

Somaclonal variation exists and has been proven to be genetically inherited. Success in somaclonal variation is genotype-dependent. Though several transformation techniques have been developed, generally speaking such techniques are highly sophisticated and depend on availability of an efficient and reproducible tissue culture system. Somaclonal variation, on the other hand, is a relatively simple technique, usable where other methods are not feasible or where resistance genes or genes of interest are not available.

Cytological Analysis

Cytological somaclone analysis gave a constant chromosome number as observed in original hexaploid wheat genotypes ($2n = 6x = 42$), whereas haploid plants had only half of the number of chromosomes ($n = 21$).

Effect of 2,4-D Concentration

Production of hybrid embryos was not wheat-genotype dependent but depended on the concentration of 2,4-D. Embryo and haploid plant numbers were severely affected by 2,4-D concentration. The higher 2,4-D concentration resulted in greater embryo and haploid plant numbers in both cultivars. Embryos were also produced when randomly selected wheat cultivars were crossed with maize and the 2,4-D concentration was 0.001%. Although embryo frequencies and haploid plant production were much lower when this lower 2,4-D concentration was used, the lower concentration is preferred as it avoids further somaclonal/gametoclonal variation, in haploid plant production.

Production of Haploids and Dihaploides Using Somaclonal Plants

Wheat and maize hybridization yielded 81 embryos when rescued 13-14 days after pollination. The seeds contained haploid embryos floating in a liquid not characteristic of endosperm. Emasculated and unpollinated florets covered with glycine bags and maintained as controls revealed no caryopsis or embryos.

Of a total of 11,624 florets of somaclonal plants crossed with maize, 8.4% produced haploid embryos across the seven wheat genotypes. Caryopsis formation with or without visible embryos is interpreted as a sign of fertilization. A high fertilization frequency does not necessarily correspond to a high embryo frequency (Comeau *et al.*, 1998). In the case of cv. 'IAPAR 40', for example, percentage of caryopsis production per pollinated florets was about 14.8%, whereas percentage of visible embryos in relation to caryopsis number was only about 10.6%. This is because clusters of some tiny embryo-like structures in the form of calli, lacked a well-defined scutellum, and hence did not produce any haploid plants on the medium even after a prolonged incubation period of 90 days. Some clearly visible embryos also failed to germinate and produced calli.

To enhance plant regeneration, Zang *et al.* (1996) cultured the pre-embryos first in an MS medium containing 2 mg 2,4-D/l to induce callus and later transferred the calli to a regeneration medium (MS medium supplemented with 2 mg kenetin, 0.5 mg IAA, 0.2 mg 2,4-D, 200 mg glutamine, and 400 mg casein/l) to induce totipotent plant formation. This procedure would undoubtedly increase the probability of somaclonal and/or gametoclonal variation of any kind in dihaploid plants due to mutations provoked by constant use of 2,4-D, i.e., twice in the hybridization process with maize and twice in the callus and the regeneration media.

Riera-Lizarazu *et al.* (1992), used a very high concentration of 2,4-D (100 mg/l) in detached tiller culture. They believed that tissue-culture-associated somaclonal variation could be avoided by polyhaploid production involving wheat x maize hybridization. However, no evidence for this hypothesis has so far been encountered in the literature. We suspect the presence of some somaclonal variation, especially when pre-embryos are forced to produce calli and later to regenerate totipotent plants. As stated earlier, we believe that the lower concentration of 2,4-D (0.001%) must have minimized the secondary somaclonal/gametoclonal variation.

Production percentage of well-developed embryos in relation to number of florets pollinated obtained was somewhat lower when compared to results obtained by other authors, possibly because we used a five-times lower concentration of 2,4-D (0.001%) for spike injection during the hybridization process. Haploid plant production frequencies of 1-4% have been considered acceptable but with the detached tiller method it could reach as high as 25%.

The procedure using wheat x maize hybrids allows complete homozygosity within one generation and facilitates the somaclonal variant selection process. Gametoclonal variation is known to occur in doubled haploids. According to Zang *et al.* (1996), haploid embryo production frequency and plant regeneration is affected significantly by maize genotypes, but not by wheat genotypes. Although no haploid embryo was produced in cv. 'BR 18', in all the other cultivars and in the somaclones, hybrid embryos and haploids were produced, thereby indicating that the wheat genotype did not affect haploid plant production. Failure to produce haploid embryos in 'BR 18' was probably due to the low 2,4-D concentration used in our study. Zang *et al.* (1996) reported that the concentration of 2,4-D also affected embryo size.

Chapter 4

Respiration and Lipid Metabolism

Although it is possible to measure respiration by intact tissues it is often useful to be able to isolate uncontaminated, intact and functional mitochondria. Here, we will consider some of the principles for isolating cell organelles. This discussion is not meant to be exhaustive nor will it be a cookbook with recipes to be followed step-by-step.

Choice of Material

If the aim of the experiment is to investigate a specific physiological problem, then the plant material is often well-defined. However, in many cases there is some flexibility in the choice of material and it is therefore wise to follow certain rules of thumb:

- Choose a tissue that is easy to obtain in relatively large amounts (at least 100–200 g fresh weight), such as tubers or hypocotyls.
- Choose a tissue for which a good protocol is available. It can take quite a lot of time and effort to optimize an isolation protocol.

Note that the isolation of mitochondria from green tissues presents special problems because of the large amount of chloroplasts. Roots are a notoriously difficult tissue for isolation of mitochondria.

Disruption of the Tissue

The aim is to rupture as many of the cells in the tissue as possible in order to release their cytoplasm with the mitochondria. Whereas many mammalian tissues are easy to homogenize, plant tissues are much tougher mainly because of the cell walls. There are a number of standard homogenizers, including mixers with rotating knives or razor blades, juice extractors, ball mills, and mortar and pestle; the most suitable method for the specific plant material should be chosen. This may involve some initial optimization experiments.

Homogenization Medium

The medium for homogenizing rat liver only contains an osmoticum and a buffer. That will not work with plant tissues because of the disruption of the large central vacuole (absent in mammalian cells), which releases its fairly acidic content often containing phenols and other potentially harmful substances. Let us consider the components of the optimal homogenization medium for isolating intact and functional organelles from plant material, one by one.

Osmoticum. This is required to prevent the mitochondria from losing their matrix content because of swelling and bursting. For the isolation of plant mitochondria, there is a tradition of using sucrose or mannitol at 0.2–0.5 M; whereas 0.15 M KCl is standard for the isolation of mammalian mitochondria. In neither case does this mimic precisely the conditions in the cytosol, but it works.

Buffer. The pH in the cytosol of plant cells is around 7.5, and the homogenate should be buffered at pH 7–8. A buffer should therefore be included and a plethora of different buffers are available. Choose an inexpensive one with a pKa of 7–8 (e.g., 4-morpholinopropane sulfonic acid = MOPS) and be sure to use enough to counterbalance the acidic vacuolar content. The tissue type, the tissue/medium ratio, and the buffer concentration determine whether the acidic vacuolar content is properly neutralized. Obviously, special tissues like lemons or CAM plants harvested late at night or in the early morning, require special treatment!

Metal ion chelators. EDTA is usually included to bind mainly Ca^{2+} and Mg^{2+}, which may otherwise activate proteolytic enzymes. Antioxidants. Antioxidants (reductants) such as ascorbic acid, dithiothreitol, or cysteine are used mainly to protect sulfhydryl groups on proteins against oxidation.

Bovine serum albumin (BSA). This protein from the blood of cows is usually included to protect the mitochondria: BSA binds fatty acids (its function in the blood) released by the action of phospholipases on phospholipids. BSA may also bind phenolics that are released from the vacuole. Both of these groups of compounds can otherwise be troublesome later. In addition, BSA may prevent proteases from degrading mitochondrial proteins by serving as an alternative substrate present at high concentrations.

Polyvinylpyrrolidone (PVP). A polymer that is often included, especially when using leaf material, to adsorb phenolic compounds.

Filtration

To remove larger tissue fragments, cell walls, and often starch grains, the homogenate is always filtered, usually through a nylon net with a mesh size of 30–100 μm.

Purification of Mitochondria

The first step of the purification is usually differential centrifugation. This means that the filtered homogenate is centrifuged first at low speed for a short time (e.g., 3000 g for 5–10 min.) and the pellet (containing cell wall fragments, starch grains, nuclei, intact plastids) is discarded. The supernatant is then centrifuged faster and for a longer time (e.g., 10,000 g for 10–20 min.) and the pellet (termed *crude mitochondria*) is collected. It contains intact and damaged mitochondria as well as peroxisomes and plastid membranes. The supernatant is discarded.

As soon as the mitochondria have been pelleted and thus removed from potentially damaging soluble substances in the homogenate, the medium can be greatly simplified. Usually the so-called wash medium contains only osmoticum and weak buffer (no need for a strong buffer since there is no vacuole content to neutralize).

However, it may be wise to include EDTA to bind divalent cations as well as BSA to adsorb any fatty acids released by the action of phospholipases. The resuspended crude mitochondria are often diluted to 50–100 ml with wash medium and subjected to another round of slow and fast centrifugation to remove more contaminants.

The aim of the next step is to separate the intact mitochondria from the contaminants. This usually involves a density gradient centrifugation with Percoll either alone (e.g., Struglics et al. 1993) or together with other dense materials (e.g., Day et al. 1985). Percoll (colloidal silica coated with PVP) is inert and osmotically inactive. It has the further advantage that a Percoll gradient is self-generating because the Percoll microspheres pellet very slowly during the centrifugation.

The crude mitochondria are layered on top of the wash medium containing the appropriate concentration of Percoll (usually 25–30%). If the Percoll concentration and the speed and duration of the centrifugation have been chosen correctly, the intact mitochondria will form a distinct band, well away from the contaminants. This band can be collected, washed free of Percoll by repeated dilution and centrifugation, and either used immediately, or frozen with 5% (v/v) dimethylsulfoxide (to preserve integrity) in liquid nitrogen and stored at -80°C until use. This preparation can be stable for months, retaining the ability to form a proton electrochemical gradient over the inner membrane.

The yield is 1–50 mg mitochondrial protein per 100 g fresh weight, depending on the tissue. This represents only a small fraction of the total amount of mitochondria in the tissue. The purification of mitochondria from green leaves presents special problems and a good protocol is described in Day et al. (1985).

There is, at present, no good protocol for the isolation of both pure and functional mitochondria from Arabidopsis leaves (Hausmann et al. 2003; Keech et al. 2005), but a protocol for the purification of mitochondria from Arabidopsis cell cultures is given in Kruft et al. (2001). Finally, Boutry et al. (1984) demonstrate how purified mitochondria can be isolated from small amounts of plant tissue (down to 0.5 g fresh weight).

Mitochondria are rod-shaped organelles that can be considered the power generators of the cell, converting oxygen and nutrients into adenosine triphosphate (ATP). ATP is the chemical energy "currency" of the cell that powers the cell's metabolic activities. This process is called aerobic respiration and is the reason animals breathe oxygen. Without mitochondria (singular, mitochondrion), higher animals would likely not exist because their cells would only be able to obtain energy from anaerobic respiration (in the absence of oxygen), a process much less efficient than aerobic respiration. In fact, mitochondria enable cells to produce 15 times more ATP than they could otherwise, and complex animals, like humans, need large amounts of energy in order to survive.

The number of mitochondria present in a cell depends upon the metabolic requirements of that cell, and may range from a single large mitochondrion to thousands of the organelles. Mitochondria, which are found in nearly all eukaryotes, including plants, animals, fungi, and protists, are large enough to be observed with a light microscope and were first discovered in the 1800s.

The name of the organelles was coined to reflect the way they looked to the first scientists to observe them, stemming from the Greek words for "thread" and "granule." For many years after their discovery, mitochondria were commonly believed to transmit hereditary information. It was not until the mid-1950s when a method for isolating the organelles intact was developed that the modern understanding of mitochondrial function was worked out.

The elaborate structure of a mitochondrion is very important to the functioning of the. Two specialized membranes encircle each mitochondrion present in a cell, dividing the organelle into a narrow intermembrane space and a much larger internal matrix, each of which contains highly specialized proteins.

The outer membrane of a mitochondrion contains many channels formed by the protein porin and acts like a sieve, filtering out molecules that are too big. Similarly, the inner membrane, which is highly convoluted so that a large number of infoldings called cristae are formed, also allows only certain molecules to pass through it and is much more selective than the outer membrane.

To make certain that only those materials essential to the matrix are allowed into it, the inner membrane utilizes a group of transport proteins that will only transport the correct molecules. Together, the various compartments of a mitochondrion are able to work in harmony to generate ATP in a complex multi-step process.

Mitochondria are generally oblong organelles, which range in size between 1 and 10 micrometers in length, and occur in numbers that directly correlate with the cell's level of metabolic activity. The organelles are quite flexible, however, and time-lapse studies of living cells have demonstrated that mitochondria change shape rapidly and move about in the cell almost constantly. Movements of the organelles appear to be linked in some way to the microtubules present in the cell, and are probably transported along the network with motor proteins. Consequently, mitochondria may be organized into lengthy travelling chains, packed tightly into relatively stable groups, or appear in many other formations based upon the particular needs of the cell and the characteristics of its microtubular network.

The mitochondrion is different from most other organelles because it has its own circular DNA (similar to the DNA of prokaryotes) and reproduces independently of the cell in which it is found; an apparent case of endosymbiosis. Scientists hypothesize that millions of years ago small, free-living prokaryotes were engulfed, but not consumed, by larger prokaryotes, perhaps because they were able to resist the digestive enzymes of the host organism. The two organisms developed a symbiotic relationship over time, the larger organism providing the smaller with ample nutrients and the smaller organism providing ATP molecules to the larger one. Eventually, according to this view, the larger organism developed into the eukaryotic cell and the smaller organism into the mitochondrion.

Mitochondrial DNA is localized to the matrix, which also contains a host of enzymes, as well as ribosomes for protein synthesis. Many of the critical metabolic steps of cellular respiration are catalysed by enzymes that are able to diffuse through the mitochondrial matrix. The other proteins involved in respiration, including the enzyme that generates ATP, are embedded within the mitochondrial inner membrane. Infolding of the cristae dramatically increases the surface area available for hosting the enzymes responsible for cellular respiration.

Methods for Establishing Purity, Intactness, and Functionality of the Fractions

It is often important to determine the degree of purity of the mitochondria and their subfractions. The intactness of the mitochondria is usually established with enzyme assays that take advantage of the fact that an intact membrane is not permeable to certain molecules. For instance, the outer mitochondrial membrane is impermeable to cytochrome *c*, a 12.5 kDa protein that is too large to pass through the outer membrane pores, and NADH cannot pass through an intact inner membrane.

Functionality is usually tested using an oxygen electrode. If the mitochondria show high rates of oxygen consumption, high respiratory control ratios (>3), and ADP:O ratios close to the theoretical values they are in very good shape. Note that, for mitochondria with an appreciable alternative oxidase activity (measured in the presence of the cytochrome oxidase inhibitor, KCN), the respiratory control ratio and ADP:O ratio should be measured in the presence of an alternative oxidase inhibitor such as salicylhydroxamic acid.

Mitochondrial Activities can be Determined in Situ

It is relatively straightforward to measure the activity of enzymes in the soluble matrix fraction isolated from mitochondria. However, this fraction is very diluted compared to the matrix inside intact mitochondria (300–500 mg protein/ml!) and protein–protein interactions may have been lost. When measuring the activity of enzymes in fractions containing membranes—intact mitochondria and submitochondrial particles—a detergent can often be used to remove the permeability barrier that the inner membrane constitutes.

However, it is also possible to use the ionophore alamethicin to permeabilize the inner membrane to molecules <1000 Da, which makes it possible to measure the activity of matrix and inner membrane enzymes in their native proteinaceous environment. Using alamethicin, it is even possible to measure the activity of cytosolic and mitochondrial enzymes inside intact cells since the ionophore also permeabilizes the plasma membrane.

The Q-Cycle Explains How Protons Can Be Pumped across the Inner Mitochondrial Membrane

The electron transport chain (ETC) in the mitochondrial inner membrane of most eukaryotes consists of four complexes, complex I to complex IV. Complexes I, III and IV pump protons over the inner mitochondrial membrane when transporting electrons. For each electron pair (corresponding to one NADH or one oxygen atom), four protons are pumped by complex I, four by complex III, and two by complex IV. The detailed three-dimensional structures have been determined for the mammalian complexes III and IV, but not for complex I. The structural information has improved the understanding of the electron paths of complexes III and IV, but the proton-pumping function of the latter is still not fully understood. For complex I, neither electron nor proton paths are known in any detail.

Ubiquinone (Q), also called coenzyme Q, is a mobile electron and proton carrier in the inner mitochondrial membrane, structurally and functionally similar to plastoquinone in the thylakoid membrane of plastids. The ability of Q to carry both electrons and protons—and that it can accept and donate them step-wise—is utilized in electron transport and proton pumping (Nicholls and Ferguson 2002).

To be reduced, it can receive two electrons from complex I or succinate dehydrogenase and take up two protons from the matrix, forming QH_2. However, at the oxidizing side (complex III), there is actually both oxidation and reduction taking place. A QH_2 will be oxidized to Q at the outer Q-site, releasing two electrons. One of these electrons goes to cytochrome *c*, whereas the other is recycled to reduce an oxidized Q in the inner Q-site—one step to becoming a semiquinone.

The next time a QH_2 is oxidized at the outer Q-site, the semiquinone in the inner site will receive a second electron, becoming a QH_2. So, from two QH_2, we have made one Q and one QH_2, which again can be oxidized. In each cycle (i.e., for each electron passing to cytochrome *c*), two protons are taken up from the matrix and two protons released to the intermembrane space. So, the Q-cycle mediates the pumping of an extra proton per electron transported, increasing energy conservation. If we look at the whole reaction from succinate to cytochrome *c*, we see that for each succinate, four protons are transported from the matrix to the intermembrane space. However, at the same time, the two electrons are also transported out. So, this corresponds to the net pumping out of two positively charged species. Since, energetically, the mitochondrial membrane potential (which determines the energetic cost of transporting a positive charge out) is much larger than the pH gradient (which determines the cost for electroneutral export of a proton), the complex III proton pumping is less energy-demanding than, for example, complex I, which pumps four protons in their completeness.

Multiple Energy-Conservation Bypasses in Oxidative Phosphorylation of Plant Mitochondria

The electron transport chain (ETC) in the mitochondrial inner membrane of most eukaryotes consists of four large protein complexes, complex I to complex IV. The electron transport activity of complexes I, III and IV (the latter two, often jointly called the cytochrome pathway) is coupled to extrusion of protons across the inner mitochondria.

The electrochemical proton gradient that is formed and maintained in the process is used by the ATP synthase (also called complex V) to make ATP . The plant ETC, however, is highly branched. As a consequence, the plant oxidative phosphorylation system has several alternative pathways of electron transport via type II NAD(P)H dehydrogenases and the alternative oxidase (AOX), and additionally, an alternative proton transporter called the uncoupling protein (UCP).

These proteins mediate bypasses around the proton-translocating multi-protein complexes (I–V): type II NAD(P)H dehydrogenases bypass complex I, AOX bypasses complexes III and IV, and UCP bypasses the ATP synthase (Siedow and Umbach 1995; Vanlerberghe and McIntosh 1997; Møller 2001; Vercesi 2001; Millenaar and Lambers 2003; Fernie et al. 2004; Rasmusson et al. 2004).

The type II NAD(P)H dehydrogenases and the AOX transport electrons without pumping protons, whereas the UCP allows proton flow from the intermembrane space to the matrix without ATP synthesis. The consequence of this is that flux through the energy conservation bypasses will neither contribute to respiratory ATP production, nor will they be controlled by cellular adenylate status. The special features of these energy bypass proteins and discuss the consequences of their existence to the plant.

NAD(P)H Dehydrogenases in Plant Mitochondria

Except for yeast, the ETC of all eukaryotes—including plants—contains a proton-pumping NADH dehydrogenase, complex I, which is inhibited specifically by rotenone. Additionally, plant, fungal and protist mitochondria contain different setups of rotenone-insensitive enzymes capable of oxidizing NADH and/or NAD(P)H and passing the electrons on to ubiquinone.

These NAD(P)H dehydrogenases are classified as type II, whereas complex I belongs to type I. The inner mitochondrial membrane in plants contains such NAD(P)H dehydrogenases, facing either the matrix or the intermembrane space, thus oxidizing NAD(P)H either from the matrix or the cytoplasm.

The outer membrane of all mitochondria contains NADH-ferricyanide and NADH-cytochrome *b* reductase activities that are insensitive to rotenone and to the complex III inhibitor, antimycin A.

There are Three NAD(P)H Dehydrogenases on the Inner Matrix Surface of the Inner Membrane

Much of the information about the number of matrix-facing NAD(P)H dehydrogenases has come from experiments using inside-out inner membrane vesicles. Such vesicles have the matrix surface facing the medium and the external NAD(P)H dehydrogenases hidden inside the vesicles and unable to interfere with the assays.

Recently, this method has been complemented with a technique where one uses a peptide called alamethicin to make a channel for the substrate through the inner membrane. Experiments with these methods show the presence of three NAD(P)H dehydrogenase activities on the matrix side of the inner membrane:

- Complex I pumps protons, is Ca^{2+}-independent and is inhibited by rotenone and diphenylene iodonium (DPI). Complex I has a high affinity for NADH.
- ND_{in}(NADH) does not pump protons, is Ca^{2+}-independent and DPI insensitive. The enzyme has a tenfold lower affinity for NADH than complex I.
- ND_{in}(NADPH) does not pump protons, is Ca^{2+}-dependent and very sensitive to DPI.

Because of the contrasting affinities of complex I and ND_{in}(NADH) for NADH, only complex I is engaged at low matrix concentrations of NADH. However, when the NADH concentration in the matrix increases (e.g., if the ETC is restricted by lack of ADP in the cell), ND_{in}(NADH) can become engaged. When electrons are transported through the type II NAD(P)H dehydrogenases, only the two proton pumping sites of the cytochrome pathway are involved in the flow of electrons from matrix NADH to oxygen. As a consequence, the theoretical ADP/O yield is 1.5 unless the AOX or the UCP are active, in which case it will be even lower.

The External NAD(P)H Dehydrogenases are Ca^{2+}-Dependent

Plant mitochondria can oxidize cytosolic NADH and NAD(P)H directly via external dehydrogenases. The NADH and NAD(P)H dehydrogenase activities are catalyzed by separate enzymes. None of the external dehydrogenases pumps protons, which means that, at most, two sites of proton-pumping are involved when the electrons move from the dehydrogenases to oxygen. So, as for ND_{in}(NADH), the theoretical maximum ADP/O yield is 1.5.

The oxidation of exogenous NADH or NAD(P)H by most plant mitochondria requires Ca^{2+}, and as a result, is inhibited by EDTA and EGTA, chelators of di-and trivalent cations. *In vitro*, the Ca^{2+}concentration needed for half-maximal activity of NADH oxidation is 0.1–0.5 μM, depending on ionic environment and respiratory state This is in the same range as the 0.1–0.2 μM free Ca^{2+} thought to be present in the cytosol of a resting plant cell. Therefore, the external NAD(P)H oxidation may become activated by the increased Ca^{2+}-concentration associated with cellular excitation due to external conditions.

Mammalian mitochondria lack external NAD(P)H dehydrogenases. Instead, they have a Ca^{2+}-dependent glycerol-3-phosphate dehydrogenase on the outer surface of the inner membrane, which donates electrons directly to the ubiquinone pool.

The dihydroxyacetone phosphate formed is converted back into glycerol-3-phosphate by a cytosolic NAD-dependent glycerol-3-phosphate dehydrogenase. In this way, the two low-molecular-weight compounds, glycerol-3-phosphate and dihydroxyacetone phosphate, shuttle redox equivalents from the cytosol, across the outer membrane and to the inner membrane to yield the same amount of ATP as is produced by the plant mechanism. Yeast have both external NADH dehydrogenases and a glycerol-3-phosphate shuttle, and since a mitochondrial glycerol-3-phosphate dehydrogenase has been found in plant mitochondria, plants may also have both systems.

Another way to oxidize cytosolic NAD(P)H that is found in mammalian mitochondria is to use the malate/aspartate shuttle to tranfer reductant (a collective term for high-energy electrons carried by, for example, NAD(P)H, ferredoxin or an organic acid redox couple like malate/oxaloacetate) into the matrix as NADH that can be oxidized by the internal NAD(P)H dehydrogenases. However, it is not clear whether this shuttle can work for import of cytosolic reductant into plant mitochondria. In the opposite orientation, NADH is shuttled from the matrix to the cytosol via the malate/oxaloacetate shuttle. In this way, the external NADH dehydrogenase can also work as a bypass around complex I. Support for this bypass has been seen in the form of an upregulation of external NADH dehydrogenase activity in mutants where complex I or an internal type II NAD(P)H dehydrogenase has been inactivated.

The Molecular Identities of the Rotenone-Insensitive NAD(P)H Dehydrogenases

The type II NADH dehydrogenases from *E. coli* and yeast are small, single polypeptide proteins of around 50 kDa. They oxidize only NADH and contain FAD as an electron-accepting prosthetic group. Attempts to purify the type II NAD(P)H dehydrogenases have been unsuccessful in identifying the plant proteins. However, the characterization of potato homologues to the type II NADH dehydrogenases in *E. coli* and yeasts have provided candidate plant proteins to try and assign to the different reported activities.

One protein, NDA1, is internal and the other, NDB1, external of the inner mitochondrial membrane. The latter was shown to be an external calcium-dependent NAD(P)H dehydrogenase. Like a related fungal NAD(P)H dehydrogenase, NDB1 contains a motif for Ca-binding, indicating that in the activation of plant and fungal NAD(P)H dehydrogenase by Ca^{2+}, the ion binds directly to the enzyme.

In Arabidopsis, a weed for which the nuclear genome sequence is known, small families of both NDA-type and NDB-type proteins are encoded by nuclear genes. One of the NDA-type proteins has been shown to be an internal NAD(P)H dehydrogenase. Additionally, a third type of type II NAD(P)H dehydrogenase was found in Arabidopsis, the NDC1. This protein is most closely related to cyanobacterial enzymes, suggesting that the gene entered the eukaryotic domain with the plastid, later to be transferred to the nucleus.

The Alternative Oxidase

The presence of the AOX is one of the features that sets plant (and fungal) mitochondria apart from mammalian mitochondria. This enzyme has therefore received a lot of attention from plant scientists and we now have a fairly good understanding of the regulation of its structure and activity. The AOX is a quinol-oxygen oxidoreductase and it does not pump protons. The oxygenase transfers electrons from ubiquinol (the reduced form of ubiquinone) to oxygen and generates water as the end-product of the reaction, so four electrons must be transferred to oxygen.

The precise mechanism is not known, but the protein has been shown to contain a catalytical di-iron centre. The functional form of the enzyme is a dimer with the two polypeptides either covalently or non-covalently bound to each other. The predicted structure places the protein embedded into the inner leaflet of the inner mitochondrial membrane, with the active site and the regulatory cysteines exposed to the matrix.

When electrons from NADH oxidation flow through the AOX, at least two of the three sites of proton translocation are bypassed, so energy conservation in the form of ATP is much smaller when the AOX is active. The activity of the AOX must therefore be carefully regulated. There are often several AOX isoforms in each plant species—for example, five genes encoding AOX proteins are present in Arabidopsis. The relative amounts of three isoforms of AOX vary with development, and cellular redox status, as well as external factors.

The AOX is present in the inner membrane as a dimer in which the two subunits can be covalently linked through a disulfide bridge between cysteine residues. The reduced non-covalently linked dimer is the active form, and this form appears to be the predominant one *in vivo*. It was recently shown that a mitochondrial thioredoxin system regulates AOX. Thioredoxins regulate enzymes by reducing disulfide bridges, and it is likely that the regulation involves the redox-active cysteines.

The presence of thioredoxin regulation connects the AOX activation level to the mitochondrial NAD(P)H pool and the oxidation of citric acid cycle metabolites by NADP-utilizing enzymes. The activity of AOX is also stimulated by 2-oxo organic acids—in particular, pyruvate, which activates by interacting with the same cysteine that is involved in dimerization.

This activation may be a feed-forward mechanism that upregulates the capacity of the ETC when the glycolytic end-product is available. Both the cytochrome pathway and the AOX pathway use ubiquinol as the substrate. In the absence of pyruvate, the cytochrome pathway is saturated at 40% Q_{red}/Q_{tot} (40% of the UQ is reduced), at which point the AOX pathway is not engaged at all.

The AOX pathway starts accepting electrons at 50% reduction and is not fully saturated at 80%.

In the presence of pyruvate, the AOX becomes engaged at a much lower Q_{red}/Q_{tot} (20%), whereas the cytochrome pathway is unaffected. It is puzzling, however, that the normal *in vivo* concentration of pyruvate appears to be higher than the concentration of pyruvate required to give maximal stimulation of AOX activity *in vitro*. It thus appears that there is little regulation of AOX activity via pyruvate under physiological conditions.

AOX Activity Can Be Measured In Vivo

For many years, benzhydroxamic acid derivatives,, were used to quantify the capacity and activity of the AOX pathway in both isolated mitochondria and intact tissues. AOX capacity was determined from the rate of respiration in the presence of KCN (which blocks the cytochrome pathway), and corrected for residual respiration in the presence of both KCN and SHAM.

AOX *in vivo* activity was determined from the decrease in the rate of respiration when SHAM was added in the absence of KCN. The assumption was that the cytochrome pathway was always used first and that the AOX would be engaged only when the cytochrome pathway was saturated. However, as we have seen above, this assumption is not correct. The consequence of this is that *in vivo* activity of AOX cannot be measured using inhibitors alone.

A better method for measuring AOX *in vivo* activity is now available. It makes use of the fact that cytochrome *c* oxidase and AOX discriminates differently between ^{18}O relative to ^{16}O.

The so-called oxygen discrimination of the sample (tissue or mitochondria) can be measured by mass spectrometry: (1) with KCN to resolve the oxygen discrimination by AOX; (2) with SHAM to resolve the oxygen discrimination by cytochrome *c* oxidase; and (3) without inhibitors to resolve the discrimination value resulting from the activity of both enzymes. From these values, the percentage contribution by AOX and cytochrome *c* oxidase to the total uninhibited rate of respiration can be calculated. An example of the use of the oxygen discrimination technique with an intact tissue is given below. A remaining problem in the estimation of AOX activity *in vivo* is that measurements can only be made in darkness (to avoid interference from photosynthetic oxygen exchange).

Plant Mitochondria also Contain an Uncoupling Protein

Mammalian mitochondria contain neither type II NAD(P)H dehydrogenases nor AOX. However, they have a UCP that increases the proton permeability of the inner mitochondrial membrane and, in that way, dissipates the proton gradient. This is another mechanism for reducing the ATP production and increasing heat production, for example, for thermoregulation.

Surprisingly, plant mitochondria also contain a UCP. We do not know for certain why plant mitochondria should require two different mechanisms for achieving the same end result. The UCP is related to anion carriers. Its *in vitro* transport activity is dependent on free fatty acids and is inhibited by nucleotides like ATP and GTP but stimulated by superoxide. The latter result indicates that the UCP may protect against oxidative stress.

The Physiological Impact of Energy Bypass Proteins

The respiratory rates of some thermogenic flowers are the highest among plants and, in fact, exceed even those of warm-blooded animals. The external NADH dehydrogenase, AOX and UCP appear particularly active in thermogenic flowers where they are primarily responsible for heat production. At the same time it is clear that the respiratory rate of most plant tissues is insufficient to generate an appreciable increase in tissue temperature.

AOX is up-regulated by cold stress, although this is clearly not to heat the tissue. However, during cold stress, reactive oxygen species (ROS) are formed, and also other stress treatments that lead to elevated ROS will induce AOX. ROS (e.g., superoxide and hydrogen peroxide) are formed as by-products of electron transport under aerobic conditions.

ROS can cause damage to proteins, lipids, and DNA and the cell must therefore limit their formation. AOX helps prevent overreduction of the ETC and thus lowers ROS production. Supporting this, Maxwell et al. (1999) demonstrated the importance of the AOX for limiting ROS formation in living plant cells.

Overexpression of AOX lowered the steady state level of ROS, whereas suppression increased it. This result also points to the mitochondrion as an important site of ROS generation in plant cells. With the recent discovery of ROS stimulation of plant UCP and the increased ROS resistance in plants overexpressing UCP, the picture has become more complex. By decreasing the resistance for electron transport through the proton-pumping complexes, UCP activity may, like AOX, lead to decreased overreduction and ROS formation. Another factor speaking in favour of UCP having a role in ROS protection is that the AOX is inhibited by an ROS product.

Both type II NAD(P)H dehydrogenases and AOX may act as reductant overflow enzymes by reoxidizing surplus NADH to oxygen. This can prevent inhibition of the citric acid cycle under conditions in which the ADP concentration is low, and thus the proton-pumping complexes are restricted by a high membrane potential. In this way, the citric acid cycle can supply carbon skeletons for biosynthesis, even if the ATP level in the cell is high.

The concept of overflow is also relevant for photorespiratory conditions, under which there is a massive flow of glycine that has to be oxidized by the mitochondria. As much as half of the NADH produced in this process is probably exported, but the other half has to be oxidized promptly. There are several reasons to believe that a reductant overflow via energy bypass proteins is involved.

The ND_{in}(NADH) and AOX capacity in leaf mitochondria is higher than in mitochondria from heterotrophic tissue, and both proteins are down-regulated by dark treatment. Consistent with these observations, AOX activity *in vivo* is up-regulated by light (Ribas-Carbo et al. 2000), AOX capacity is induced by greening of etiolated tissue, and expression of the *nda*1 gene [encoding an internal NAD(P)H dehydrogenase] in potato and Arabidopsis is light-dependent. Mutant tobacco plants, lacking the proton-pumping complex I can still survive, apparently aided by the action of ND_{in}(NADH), which oxidizes matrix NADH with a lower ATP yield.

These plants, however, show decreased photosynthetic rates, indicating that the ND_{in}(NADH) cannot fully compensate for the missing complex I. Also, external NAD(P)H dehydrogenases have been suggested to support photosynthesis. These enzymes may reoxidise NAD(P)H that is shuttled out of the chloroplast in order to balance the NAD(P)H to ATP ratio in the stroma. In summary, there are many indications that mitochondria—especially the energy bypass proteins—have roles in supporting photosynthesis, for example by oxidizing excess NAD(P)H.

The AOX pathway is involved in the acidification/deacidification cycle of CAM plants. CAM plants fix CO_2 in the form of malate during the night. This is done by cytosolic PEP carboxylase, and the malate produced is stored in the vacuole.

The CO_2 fixed is released during the daytime by malic enzymes, located in cytosol and mitochondria. Robinson et al. (1992) used the oxygen discrimination technique to measure the contribution of the AOX pathway to whole-leaf respiration in the CAM-plant, *Kalanchoë daigremontiana.*

They found that the respiratory rate of the leaves increased by 35% in the early hours of the day. This increase was due to increased AOX activity. A more detailed time course showed that the AOX activity increased two-to threefold during the early day at the same time as the malate was metabolized. As soon as the malate had disappeared by mid-day, the AOX activity returned to the low level observed during the night. A reasonable conclusion is that a substantial amount of the malate from the vacuole is oxidized via malic enzyme in the mitochondria. This generates NADH in the matrix, which is oxidized to a large extent via the AOX. The released CO_2 can then be refixed by the Calvin cycle in the chloroplast.

F_oF_1-ATP Synthases: The Worlds Smallest Rotary Motors

F_oF_1-ATP synthases (also called F-ATPases) are present in the inner membranes of mitochondria, chloroplasts, and bacteria. These large, multisubunit enzymes consist of a water-soluble catalytic complex (F_1) attached to an integral membrane protein complex (F_o) that transports protons across the membrane. The F_1 complex is composed of at least five different types of subunits (three α, three β, one γ, one δ, and one å).

When the F_1 complex is dissociated from the membrane, it is active as an ATPase. In fact, under the appropriate conditions the intact F_oF_1-ATP synthase can run in reverse and act as a proton pump, using the energy of ATP hydrolysis to move H^+ across the membrane. Our understanding of how ATP is synthesized was advanced by Paul Boyer at the University of California, Los Angeles, who proposed the binding-change mechanism for catalysis by F-ATPases. From the detailed understanding of how the ATPases function, he proposed that the binding-change mechanism for ATP synthesis contains three important components:

1. The major energy-requiring step is not the synthesis of ATP from ADP and P_i, but the release of ATP from the enzyme.
2. Substrate is bound and products are released at three separate but interacting catalytic sites, corresponding to the three catalytic subunits (b subunits). Each catalytic site can exist in one of three conformations: tight, loose, or open.
3. The binding changes are coupled to proton transport by rotation of the g subunit. That is, the flow of protons down their electrochemical gradient through the F_o complex causes the g subunit to rotate.

Rotation of the γ subunit then brings about the conformational changes in the catalytic complex that allow the release of ATP from the enzyme, and the reaction is driven forward. The reverse occurs when the enzyme functions as an ATP-driven proton pump.

The first two predictions of the binding-change model are supported by many lines of evidence, mainly kinetic studies, and are now generally accepted. However, the prediction of a rotary mechanism for coupling proton flow to ATP synthesis has been more difficult to demonstrate. Two major breakthroughs have led to confirmation of the third prediction, as well.

The first breakthrough was the determination of the crystal structure of the F_1-part of bovine mitochondrial ATPase by the laboratory of John Walker in Cambridge, England. The crystal structure showed that the three catalytic β subunits differ in their conformations and in the nucleotide bound to them, consistent with the binding-change mechanism.

Even more exciting was the discovery that the γ subunit is inserted like a shaft through the centre of the catalytic complex, which consists of three α subunits and three β subunits arranged alternately in a doughnutlike structure. Moreover, the interface between the γ subunit and the α and β subunits is highly hydrophobic.

The hydrophobicity of the interface minimizes the interactions between the subunits, consistent with the rotation of the γ subunit within the hole formed by the catalytic complex. In other words, the γ subunit looks like a molecular bearing lubricated by a hydrophobic interface. Although many questions were answered by elucidation of the crystal structure of the F_1-ATP synthase, the rotational model cannot be tested by a static "snapshot" of the enzyme. Definitive demonstration of rotation requires a video recording of the spinning of the enzyme in real time. But although they are large for proteins, F_1-ATP synthases are still far too small to be visualized in a light microscope.

To visualize the rotation of the enzyme, Masasuke Yoshida and his colleagues at the Tokyo Institute of Technology came up with an ingenious method to make the enzyme much larger than it is. As Figure below shows, they attached an actin filament labelled with a fluorescent dye to the base of the γ subunit using another protein as a "glue." They then attached the F_1 complex upside down to a glass surface.

If the γ subunit rotates with respect to the catalytic complex, the actin filament should swing around with it. Since the filament is very long compared to the ATP synthase (about 1 μm), its rotation should be visible in a fluorescence microscope. In other words, the fluorescently tagged actin filament, which is large enough to visualize in a light microscope, reports the rotation of the γ subunit.

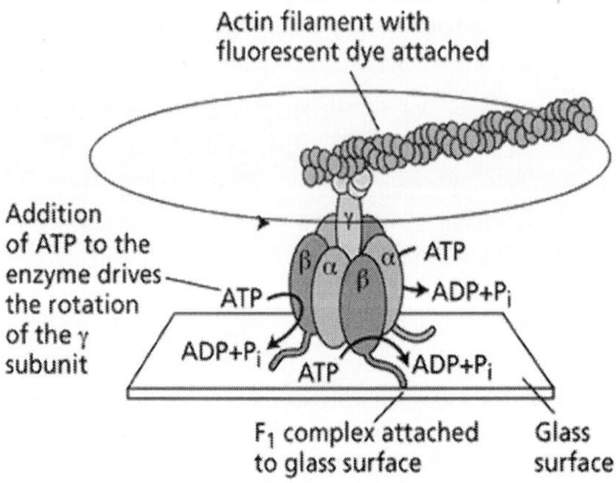

Figure: *A method for visualizing rotation of the g subunit. A fluorescently labelled actin filament was attached to one protruding end of the g subunit. The F_1 complex was then attached upside down to a coverslip. When ATP was added to the coverslip, the actin filament rotated. (After Noji et al. 1997.)*

The results were spectacular! When ATP was added to the modified enzyme, the actin filaments were seen to swing around in a circle at as much as 4 revolutions per second in a fluorescence microscope. To give some idea of scale, if you were a γ subunit, this rotation rate would be equivalent to swinging a several hundred meter-long rod around your head at 4 revolutions per second in water! However, the measured velocity is undoubtedly an underestimation of the actual velocity in vivo, because of the enormous torque required to swing such a large mass. Demonstration of the rotary motion of the γ subunit made it possible to put together a model of how the ATP synthase works.

For their contributions to elucidation of the mechanism of ATP synthesis, Paul Boyer and John Walker shared half the Nobel prize in physiology or medicine in 1997. The other half went to Jens Skou for his pioneering work on the K^+,Na^+-ATP synthase, the mammalian counterpart of the plant plasma membrane H^+-ATP synthase.

Transport Into and Out of Plant Mitochondria

Plant mitochondria must exchange metabolites and information (signal transduction) with the cytosol. To give some central examples, the end products of glycolysis, pyruvate and malate, need to be taken up, and citric acid cycle intermediates exported to be used as building blocks in biosynthetic processes.

Likewise, ADP and P_i must be taken up and ATP exported for the mitochondrion to fulfil its role as the powerhouse of the cell. To carry out the transport of this large number of molecules, plant mitochondria possess a family of mitochondrial carrier proteins. These are transmembrane proteins with a common general structure and they typically transport relatively small charged molecules, like metabolites. In Arabidopsis there are almost 60 members, most of which are not functionally characterized, giving a hint to the complexity of the mitochondrial transport processes. However, mitochondria exchange more than metabolites, and therefore also other types of transporters are involved.

The outer membrane is permeable to molecules of less than 10 kDa, (i.e., all non-polymeric compounds). The permeability results from the presence of pores, which appear to be open in most *in vitro* experiments. Recent studies in animals have characterized pore-forming proteins in the mitochondrial outer membrane, called porins, which appear to regulate transport across the membrane.

Plant mitochondria also have porins but their characterization awaits further research. In our discussion here we will focus on the regulation of mitochondrial transport across the inner mitochondrial membrane. We will first consider the relatively few molecules that can move across the inner membrane without the help of a specific carrier protein. We will then discuss the various metabolite transporters and, finally, briefly look at the transport of proteins and RNA.

Small, Neutral, and Hydrophobic Molecules Diffuse across the Inner Membrane

Irrespective of size, molecules that either are charged or are very hydrophilic can only cross a lipid bilayer with the help of a specific protein carrier.

Small, *neutral* molecules can diffuse across the lipid bilayer and do not require protein transporters. Among those are O_2 and CO_2, two important molecules for respiration. O_2 is needed on the inner surface of the inner membrane where the active sites of both cytochrome *c* oxidase and the alternative oxidase are located.

The CO_2, produced as an end product in the citric acid cycle, needs to diffuse out of the mitochondrion, ultimately to be lost to the surroundings or re-fixed in the chloroplasts. In addition, however, it is possible that CO_2 can also be transported out of the mitochondrion as bicarbonate (HCO_3^-). Mitochondria are osmotically active and swell in a hypotonic medium. This means that the inner membrane is also permeable to H_2O, but it is not clear whether this occurs by diffusion or if aquaporins may be involved.

For technical reasons, it is of importance that several well-known inhibitors of mitochondrial function can diffuse into mitochondria because they are uncharged and hydrophobic.

These are rotenone, the specific inhibitor of complex I, antimycin, the specific inhibitor of complex III, HCN, the protonated form of the cyanide (CN^-) inhibitor of complex IV (and other heme-containing enzymes), and oligomycin, the specific inhibitor of the F_oF_1-ATP synthase.

Transport of Inorganic Ions

Inorganic phosphate ($H_2PO_4^-$) is taken up into the matrix in exchange for a hydroxyl ion (HO^-). The energy for the uptake is supplied by the proton gradient (–pH). We will see below that the phosphate gradient, in turn, drives transport of other metabolites.

Several other inorganic ions—Mg^{2+}, Ca^{2+}, K^+, and NH_4^+—are involved in mitochondrial metabolism, but the mechanism by which they pass the inner membrane of plant mitochondria is not well understood. In mammalian mitochondria, uptake of K^+ through a K^+-specific channel, and extrusion via a K^+/H^+ antiporter, is involved in matrix volume regulation and cellular signalling (Garlid and Paucek 2003). A similar system of K^+ transport is present in plant mitochondria.

Both mammalian and plant mitochondria contain an uncoupling protein (UCP). This protein facilitates the movement of protons across the inner membrane and therefore uncouples electron transport and decreases ATP yield.

Transport of Adenylates

One of the important functions of mitochondria is to supply the cell with ATP. To do this, mitochondria have an adenine nucleotide transporter that facilitates the exchange of ADP^{3-} (inward) for ATP^{4-} (outward). This exchange is driven by the membrane potential (inside negative), since there is a net outward movement of one negative charge.

In combination with the phosphate carrier reaction, this means that the uptake of one ADP and one phosphate and the export of one ATP (i.e., the transport event connected to mitochondrial ATP synthesis) costs one vectorial proton—the electrical component for adenylate exchange and the "pH component for phosphate uptake.

Transport of Citric Acid Cycle Intermediates and Related Compounds

Pyruvate and malate, the end products of glycolysis, both have transporters in the inner membrane. Pyruvate is taken up in exchange for a hydroxyl ion, and malate^{2-} in exchange for $H_2PO_4^{2-}$.

The transporter for malate also takes succinate^{2-} and is therefore called the dicarboxylate transporter. Malate^{2-} or succinate^{2-} can, in turn, exchange for citrate^{2-} on the tricarboxylate transporter. Malate^{2-} can also exchange for 2-oxoglutarate^{2-} via a special transporter. An interesting transporter, the oxaloacetate transporter, exchanges oxaloacetate^{2-} for malate^{2-} (or citrate^{2-}), which permits the shuttling of reducing equivalents.

The joint activity of these transporters allows plant mitochondria to import or export most citric acid intermediates. Finally, the glutamate/aspartate transporter facilitates the uptake of glutamate. In mammalian mitochondria, glutamate is taken up together with a proton, and it is thus driven by both the electrical and the chemical proton gradient. A similar transporter is present in plant mitochondria, but a proton co-transport has not been shown, so the transport may be passive in plants. Recently, several additional transporters for metabolites have been identified, and it is likely that certain setups of transporters are only expressed under special developmental conditions, like the mobilization of storage nutrients in developing seedlings.

In several cases the involvement of the mitochondrion in cellular metabolism implies that certain metabolites are taken up across the inner membrane, but the transporter has not yet been described. This is the case for the massive flux of glycine into—and CO_2, serine and NH_4^+ out of—mitochondria in C_3 leaves during photorespiration This is also the case for proline, which accumulates to high concentrations in the cytosol during episodes of plant water stress, and it is then broken down during the rehydration phase by a pathway involving two mitochondrial enzymes.

Transport of Coenzymes

Coenzymes are either synthesized inside the mitochondrion or imported. There is some evidence that, in addition to ATP, at least NAD^+, thiamine pyrophosphate, and coenzyme A are imported across the inner membrane.

The final step in ascorbic acid biosynthesis has recently been shown to be on the outer surface of the inner mitochondrial membrane. In view of the accumulating evidence that ascorbate has an important function in removing ROS inside the mitochondrion, we should expect to find an ascorbate transporter in the inner mitochondrial membrane.

The situation with folate (involved in 1C transfers) is the opposite to that with ascorbate: The only site of synthesis of many of the folate metabolites in the plant cell is the mitochondrial matrix. We can therefore expect that these molecules are exported via specific transporters.

Transport of Reducing Equivalents—Metabolic Shuttles

Mammalian mitochondria cannot oxidize cytosolic NADH or NAD(P)H directly, so they make use of "shuttle mechanisms." An exchange of metabolites by inner membrane transporters takes place through these shuttles so that a reduced compound moves into the matrix in exchange for a more oxidized compound moving out into the cytosol.

In this way, reducing equivalents are transported into the matrix ultimately to be oxidized by the respiratory chain. The shuttles can also be used to transport reducing equivalents in the opposite direction. Although plant mitochondria can oxidize cytosolic NADH and NAD(P)H directly by the two external NAD(P)H dehydrogenases, they can also use these two metabolic shuttles. The malate/oxaloacetate shuttle uses malate dehydrogenase in the cytosol and in the matrix to catalyze the interconversion of malate (reduced) and oxaloacetate (oxidized).

These two compounds are exchanged on the oxaloacetate transporter. This exchange, however, is not driven by the electrochemical proton gradient across the inner membrane, so it can only move reducing equivalents from a relatively reduced compartment to a relatively oxidized compartment. Since the mitochondrial matrix is more reduced than the cytosol, the oxaloacetate transporter is likely to work primarily in the export of reducing equivalents. Under photorespiratory conditions, large amounts of glycine are oxidized in the mitochondria to give NADH while, at the same time, NADH is required in the peroxisomes to reduce hydroxypyruvate to glycerate. It is thought that as much as 50% of the matrix NADH is exported, probably via the malate/oxaloacetate shuttle.

The more complex malate/aspartate shuttle involves two matrix enzymes and two cytosolic enzymes as well as two inner membrane transporters. In mammals, it is powered by the electrochemical proton gradient in the direction of the import of reducing equivalents. However, since the glutamate/aspartate exchanger may be passive in plants, this shuttle in plants may similarly be independent of the electrochemical proton gradient. If so, it would (like the malate/oxaloacetate shuttle described above) only be able to function as an exporter of mitochondrial reductant.

Transport of Proteins—Protein Import

Plant mitochondria contain their own DNA (mtDNA) and ribosomes and are capable of synthesizing the proteins encoded in the mtDNA. However, plant mtDNA encodes only a few of the proteins found in the mitochondrion. The great majority of the mitochondrial proteins have to be imported from the cytosol where they are synthesized on free ribosomes. These are synthesized with a presequence or targeting peptide in the N-terminal end, which is recognized by receptors on the outer mitochondrial membrane.

The import of proteins requires a complex machinery with two multisubunit protein complexes—one in the outer membrane (TOM, transporter of the outer membrane) and one in the inner membrane (TIM, transporter of the inner membrane). Concomitant with, or immediately afterwards the import of the protein into the correct mitochondrial compartment, the targeting peptide is often cleaved off to yield the mature and active protein.

Transport of RNA

Mammalian mtDNA encodes the full complement of tRNA required for mitochondrial protein synthesis. In contrast, plant mtDNA only encodes some of the tRNAs required for protein synthesis within the mitochondrion. The missing tRNA must be imported by a yet-to-be-elucidated import mechanism.

Chapter 5

Stress Physiology

Stomatal Conductance and Yields of Irrigated Crops

In the last few decades, the breeding of agricultural crops for higher yields has been very successful. Breeders usually select for high-yielding genotypes empirically, without particular attention to specific plant traits that might be conducive to higher yields.

However, comparing old, low-yielding lines of any crop with advanced, high-yielding lines shows clearly that many morphological, physiological, and biochemical traits have been altered by the intense selection pressures for higher yields.

These changes indicate that selection for high-yielding genotypes has generated indirect selection pressures on the altered traits. If we exclude effects of genes that regulate the expression of two or more unrelated traits (pleiotropic genes), the study of high-yielding lines might reveal specific traits and genes associated with the higher yield. With this information, breeders could explicitly select for yield-enhancing traits to improve yields further.

Changes in plant traits associated with yield increases can be studied conveniently in a historical series of a crop. Historical series include successive commercial releases within a breeding program, with each member of the series representing an incremental increase in yield. In studies aimed at characterizing relationships between specific traits and higher yields, all the members of the series are grown together under the same conditions, and the traits of interest, such as plant architecture or photosynthetic rates, are measured in parallel with yields.

Recent studies of this type focusing on historical series of Pima cotton (*Gossypium barbadense*) and bread wheat (*Triticum aestivum*) have shown a remarkable positive correlation between yield increases and increases in stomatal conductance.

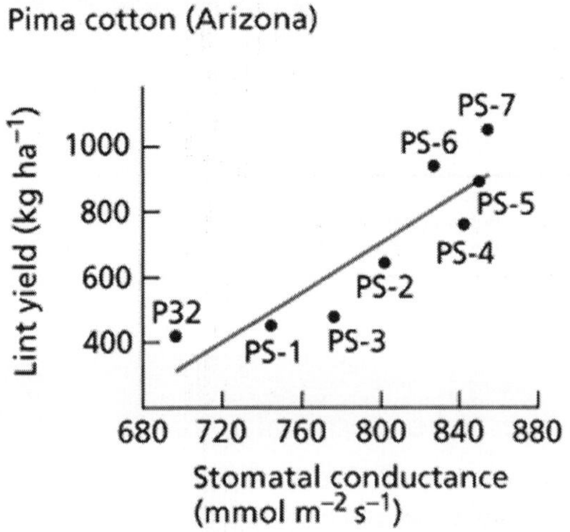

Figure: *Stomatal conductance has increased in parallel with agronomic yields in irrigated Pima cotton (*Gossypium barbadense*) selected for higher yields at a high temperature. The figure shows the relationship between lint yield and stomatal conductance in a historical series of Pima cotton grown in Arizona. The abbreviations P32 and PS-1 through PS-7 designate successive commercial releases between 1949 and 1996.*

Pima cotton is a high-quality fibre cotton grown in hot environments under intensive irrigation. The historical series used in the Pima studies included eight members, released between 1949 and 1996, and encompassing a nearly threefold increase in lint yield. The yield increase attained in each commercial release was accompanied by a corresponding increase in stomatal conductance. Stomatal conductance increased by about 30 mmol m^{-2} s^{-1} for each 100 kg ha^{-1} increase in yield.

Genetic crosses between low-and high-yielding Pima strains have shown that stomatal conductance in Pima cotton has a clear-cut genetic component. The studies with stomata of Pima cotton are the first to show a clear-cut genetic regulation of stomatal conductance proper, of the stomatal response to temperature, and of proton pumping rates of isolated guard cells.

In the absence of explicit selection for higher stomatal conductance in the Pima breeding program, what could be the nature of the indirect selection pressures that caused the increases in conductance paralleling the increases in yield? Higher stomatal conductance increases CO_2

diffusion into the leaf and favours higher photosynthetic rates. Higher photosynthetic rates could in turn favour a higher biomass and higher crop yields. Advanced Pima lines show a higher photosynthetic capacity than older, low-yielding lines, but photosynthetic rates measured in the same leaves used to measure stomatal conductance were not positively correlated with yields. Thus, higher stomatal conductance appears to favour higher yields by a mechanism not directly related to photosynthesis.

Evapotranspiration at the leaf surface lowers leaf temperature, and higher stomatal conductance enhances this leaf cooling. Optimal daytime temperatures for growth, photosynthesis, and reproduction in Pima cotton are below 30°C, while afternoon air temperature in the hot Pima-growing areas often exceeds 40°C. Evaporative cooling of the leaves thus reduces the gap between optimal growth temperatures and air temperature, and provides heat resistance. Studies with the Pima historical series showed that, because of leaf cooling, leaf temperatures were several degrees lower than air temperature, and leaf and canopy temperature were lower in the advanced, high-yielding lines than in low-yielding lines.

The same relationship among yields, stomatal conductance, and leaf temperature was found in a historical series of bread wheat grown in the warm Yaqui Valley of northwestern Mexico. These studies indicate that selection for higher yields in irrigated crops grown at high temperatures imposes indirect selection pressures for high stomatal conductance that lowers leaf temperature and appears to reduce deleterious effects of heat stress on critical flowering and fruiting stages, thus resulting in higher crop yields.

If selection for higher yields generates indirect selection pressures for higher stomatal conductance, can breeders explicitly select for higher conductance and thus obtain lines with higher yields? Studies with Pima cotton have shown that selection of high-conductance F_2 progeny from a cross between high-and low-conductance parents produces high-conductance F_4 lines with higher lint yields than low-conductance lines have. These experiments indicate that high stomatal conductance could be used as a selection trait for high yields in irrigated crops grown at high temperature.

The observed increases in stomatal conductance in crops grown at high temperature also have valuable implications for models developed to predict global climate changes, which are expected to occur because of increases in atmospheric CO_2. Under atmospheric CO_2 concentrations that are twice as high as the present ones, evapotranspiration would decrease over the continents and air temperatures would increase significantly over the tropical land masses, amplifying the changes resulting from atmospheric radiative effects.

However, these models have yet to take into consideration the effect of higher temperatures on the stomatal control of evapotranspiration. Incorporation of the stomatal response to temperature in climate change models is likely to improve the accuracy of the predictions generated by the models. These studies illustrate both the complex interactions between physical and biological factors in the biosphere, and the advantage of a thorough understanding of the physiological properties of plants for efforts aimed at increasing agronomic yields and at improving models that predict global climate changes.

The GABA Shunt Pathway Integrates Stress Metabolic Responses

Not all enzymes participating in metabolic pathways have identical temperature responses, and differential thermostabilities of enzymes may also effect specific steps in metabolism before HSPs acting as molecular chaperones can restore activity. Thus, heat stress can lead to changes in metabolism that result in the accumulation of some metabolites and the reduction of others. Such changes can dramatically alter the function of metabolic pathways and lead to imbalances that are difficult to correct. This problem can be extremely critical when the levels of regulatory metabolites are forced outside of normal ranges so that regulatory homeostasis cannot function properly to restore normal levels.

Additionally, the consumption—and production—of protons during metabolic reactions and/or changes in the activity of membrane-associated proton-transporting ATPases and other membrane-associated proton transporters that move protons out of the cytosol either into the apoplast or vacuoles, can also cause a lowering of the pH of the cytosol, leading to additional metabolic perturbations during stress. Metabolic adaptation to the stress results in the amelioration of such stress effects on metabolism.

One of the metabolic adaptations that plants make to heat stress leads to the accumulation of the nonprotein amino acid, 4-aminobutyric acid (GABA). During episodes of heat stress, GABA accumulates to levels six-to tenfold higher than in unstressed plants. GABA is made from the amino acid, L-glutamate, in a single reaction catalyzed by the enzyme glutamate decarboxylase (GAD). GAD is one of several enzymes whose activity is modulated by the calcium regulatory protein, calmodulin. In the presence of calcium ions, calmodulin is converted into an active state that binds to GAD activating the enzyme, which leads to an increase in the rate at which GABA is produced.

The precise mechanism leading to increases in cytosolic calcium have not been directly demonstrated, but studies using inhibitors of calcium ion channels implicate either the plasma membrane or tonoplast calcium channels. Alternatively, the efficiency of one or more of the CAX gene products (calcium proton antiporters) may be reduced as the proton gradient collapses during heat stress, leading to elevated levels of calcium in the cytosol. These increases, in conjunction with calmodulin, activate GAD, producing the high-temperature-induced synthesis of GABA.

GABA is metabolized inside the mitochondria by a GABA aminotransferase (GAT) that utilizes pyruvate and GABA as substrates generating alanine and succinate semialdehyde. This latter compound is then converted into succinate by succinate semialdehyde dehydrogenase (SSA-DH). The function of the mitochondrial components of this pathway leads to the reduction of pyruvate inside the mitochondria and the bypass of citric acid cycle metablism between pyruvate and succinate. This has the effect of bypassing three of the major NADH-producing reactions inside the mitochondria and constitutes a major metabolic adjustment to cellular heat stress that can dramatically alter the redox status of the cell.

Although GABA is an important signalling molcule in mammalian brain tissue, there is no evidence that it directly functions as a signalling molecule in plants. However, the synthesis of GABA consumes protons and thus the GAD-catalyzed reaction is one of several reactions involved in the regulation of cytosolic pH. The function of GAD aids the cell during the heat stress episode by maintaining cytosolic pH in a normal physiological range. Additionally, Arabidopsis mutants of SSA-DH demonstrate increased hydrogen peroxide production during episodes of heat stress and show increased programmed cell death. Thus, the function of the GABA metabolic pathway appears to modulate the oxidative status of the cell during heat stress, and may be involved in the production of a cell signal, effectively modulating PCD.

Membrane Lipids and Low Temperatures

The importance of membrane lipids for tolerance to low temperatures has been demonstrated by work with mutant and transgenic plants in which the activity of particular enzymes led to a specific change in membrane lipid composition independent of acclimation to low temperature. In one study, the genes used to modify the lipid composition of thylakoid membranes encoded glycerol-3-phosphate acyltransferases (GPATs). These enzymes select fatty acids for incorporation into PG; GPATs from different plant species have different selectivity for saturated and unsaturated fatty acids, leading to the differences in composition.

The degree of fatty acid saturation of phosphatidylglycerol (PG) is particularly important to the chilling resistance of chloroplasts. In one experiment, tobacco was transformed with an acyltransferase gene from either squash (chill sensitive) or *Arabidopsis* (chill tolerant) to modify the fatty acid saturation of PG in the thylakoid membranes of the chloroplasts. In transgenic tobacco transformed with the squash gene, which increased the proportion of high-melting-point (trans-monounsaturated and saturated) molecular species of PG, chilling tolerance decreased (88% inhibition of photosynthesis following 4 hours at 1°C). Conversely, when tobacco was transformed with the *Arabidopsis* chill-tolerant gene, the proportion of high-melting-point molecular species of PG decreased, and tolerance of chilling was greater than in wild-type plants.

Ice Formation in Higher-Plant Cells

During the initial phase of heat loss from tissue, when the temperature drops below the freezing point of the cytosol and the vacuole without ice forming, the liquid phase in the cytosol and vacuole is said to be supercooled. As the temperature drops further, ice forms in the intercellular spaces, and heat energy (0.33 kJ or 80 calories per gram) is released as a result of the latent heat of fusion of water.

At this stage, the temperature of the tissue reflects the balance between heat gain from ice formation and heat loss to the environment. As a result, when ice first forms, the temperature rises rapidly (point B in the figure), and it remains at that level until all the extracellular water is frozen (point C in the figure). When that point is reached, heat release stops, and the temperature begins to fall again. The release of heat energy during ice formation (points B to C in the figure) is the basis for the common practice of spraying crops with water during frost: As long as the water continues to freeze extracellularly, it releases heat that prevents intracellular freezing.

The formation of ice within the intercellular spaces of cells that are sensitive to freezing is not lethal, but extended exposure to freezing temperatures causes water vapour to move through the plasma membrane from the unfrozen protoplast to the cell wall, causing ice crystals to grow within the intercellular spaces. This slow dehydration concentrates solutes within the protoplast, depressing the freezing point by 2 to 3°C. As the temperature continues to drop, a second phase of release of the heat of fusion of water is detectable (points from D to E on the figure). This phase reflects a series of small freezing events: Each "spike" of heat release is thought to represent the freezing of cell protoplasts and coincides with loss of viability. The formation of ice crystals on cell walls or in the protoplasm requires the presence of ice nucleation points on which crystals can be initiated and grow.

In some species, acclimation confers an ability to suppress ice nucleation in the protoplast, allowing deep supercooling to many degrees below the freezing point without ice formation. However, deep supercooling of intracellular water has a lower limit of about –40°C, the temperature at which ice forms spontaneously. (The homogeneous ice formation temperature of pure water droplets is –38.1°C, but the presence of solutes lowers this temperature.) At or below –40°C, ice crystals form without nucleation points, and intracellular freezing and cell death are unavoidable. Species that tolerate temperatures below –40°C under natural conditions do so not by supercooling, but by tolerating gradual dehydration.

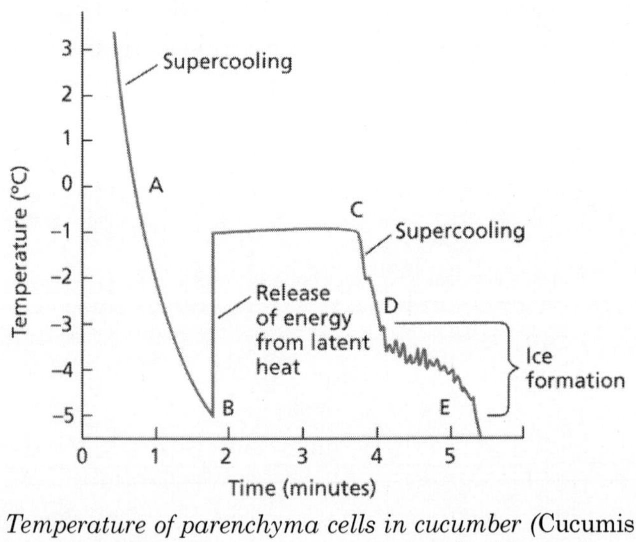

Figure: *Temperature of parenchyma cells in cucumber (*Cucumis sativus*) fruit during freezing. The temperature was recorded with an electronic device, a thermistor, inserted into a 5 × 20 mm cylinder of tissue and immersed in a coolant at –5.8°C. (A–B) Supercooling. (B–C) Release of heat during freezing in cell walls and intercellular spaces. (C–D) Supercooling. (D–E) Small heat spikes released during intracellular freezing of individual protoplasts.*

Plants produce compounds that protect cells against intracellular ice formation. Some of these compounds, such as sugars, amino acids, and other solutes, induce supercooling in the plant's tissues. Overwintering plants also produce antifreeze proteins and other low molecular weight compounds. These compounds provide freeze tolerance by inhibiting ice crystal growth and the nucleation of ice crystals.

Ca^{2+} Signalling and Activation of the Salt Overly Sensitive (SOS) Signal Pathway

Recent studies have identified components of the SOS signal pathway as genetic loci that mediate salt sensitivity.

Three genetically linked loci (*SOS1*, *SOS2* and *SOS3*) constitute determinants of a stress-signalling pathway that controls ion homeostasis and salt tolerance.

- SOS3 encodes an EF-hand type Ca^{2+} binding protein with sequence similarity to the regulatory B subunit of calcineurin (protein phosphatase 2B) and neuronal Ca^{2+} sensors. SOS3 has three EF-hand structures, and the interaction of SOS3 with SOS2 kinase leading to activation of SOS2 activity is Ca^{2+} dependent. The function of SOS3 as a salt tolerance determinant *in planta* is dependent on N-myristoylation.
- SOS2 is a serine/threonine protein kinase that is similar to yeast SNF1 and mammalian AMPK protein kinases. Although all the outputs of the SOS signal pathway have yet to be completely defined, genetic and biochemical evidence indicates that the principal effectors regulated by the pathway are transport systems involved in ion homeostasis at least some of which are the same ones discussed above.
- The plasma membrane Na^+/H^+ antiporter, SOS1, is controlled both at the transcriptional and post transcriptional levels by the SOS signalling pathway. Functional disruption of *AtHKT1* was shown to suppress the salt-sensitive phenotype of *sos3*, indicating that the SOS pathway negatively controls this Na^+-influx system. Additionally, the SOS signalling pathway negatively controls expression of vacuolar AtNHX family members.

These studies illustrate the combined use of genetic, physiological, and biochemical approaches for the elucidation of molecular determinants of ion homeostasis and their signal-mediated regulation in plants. Functional genetic studies are investigating the role of these transport proteins in tolerance to salinity stress.

Na^+ Transport across the Plasma Membrane and Vacuolar Compartmentation

The energy-dependent transport of Na^+ from the cytosol of plant cells across the plasma membrane is mediated by a Na^+/H^+ antiporter. This antiporter is known as SOS1 and it is presumably representative of a family of SOS1 genes. *Arabidopsis sos1* mutants are much more sensitive to growth inhibition by salt than are wild type plants. Interestingly, SOS1 gene expression is not regulated by NaCl stress; but it rather depends on other components of the SOS signalling pathway.

Several of the transporters that mediate Na^+ uptake into the cytosol across the plasma membrane have now been identified. In patch clamp studies with isolated membranes, sodium has been shown to compete with K^+ in inward rectifying K^+ channels. Other reports have suggested that K^+ outward-rectifying channels might facilitate Na^+ influx as well. The high affinity K^+ transporter (HKT) from wheat and the low affinity cation transporter (LCT1) from wheat facilitate Na^+ transport in patch clamp studies with isolated membranes, providing evidence to support their possible function in Na^+ uptake. Patch clamp studies have shown the presence of Ca^{2+}-insensitive-sensitive Na^+ conductances in isolated plant plasma membranes.

The Ca^{2+}-sensitive component of Na^+ uptake has been attributed to nonselective cation channels (Schachtman et al. 1997), while *in planta* studies using *Arabidopsis* mutants have directly demonstrated the function of AtHKT1 as a Ca^{2+}-insensitive Na^+-uptake system (Rus et al. 2001). These studies strongly implicate *HKT*1 in the direct control of at least some Na^+ influx. Since *HKT* transcripts are expressed predominantly in the roots, it most probably functions in the control of Na^+ flux into the xylem for export to the shoot.

Vacuolar compartmentalization of Na^+, energized by the ÄpH across the tonoplast membrane, also involves a family of Na^+/H^+ antiporters. *Arabidopsis AtNHX*1 is the prototype for such transporters. Transgenic *Arabidopsis* and tomato plants overexpressing *AtNHX*1 exhibit enhanced salt tolerance. The *Arabidopsis AtNHX*1 family include six loci showing amino acid sequence similarity and topological complementarity to *AtNHX*1, including 12 transmembrane spanning domains and the presence of amiloride binding sites. The *AtNHX*1 family of proteins plays a critical role in the vacuolar compartmentalization of Na^+, which minimizes cytotoxicity and facilitates osmotic adjustment.

Gene Transfer and Stress Tolerance

The transfer of genes suspected to mediate stress tolerance to a test plant where they are overexpressed or altered in some way has given scientists an important new tool with which to study genes involved in stress tolerance. Results of the first study in which transgenic plants were reported to show altered stress tolerance after transfer of a foreign gene appeared in 1993. The transferred gene encoded the enzyme mannitol dehydrogenase, and its expression led to the accumulation of mannitol in the transgenic plants. Transfer of other genes generated transgenic plants that accumulated other solutes, such as trehalose, proline, and fructan.

Interestingly, the amount of accumulated solutes was insufficient to osmotically balance the applied osmotic stress, yet all of the transgenic plants exhibited some increased stress tolerance. These surprising results are not fully understood, but the newly acquired solutes might protect the transgenic plant against free radicals generated in response to stress, rather than function as osmotic-balancing solutes. Improved resistance to drought, salinity, and extreme temperatures has been observed in transgenic plants that express/overexpress genes regulating osmolytes, specific proteins, antioxidants, ion homeostasis, transcription factors, and membrane composition.

Plant Physiology

Plant physiology is a subdiscipline of botany concerned with the function, or physiology, of plants. Closely related fields include plant morphology (structure of plants), plant ecology (interactions with the environment), phytochemistry (biochemistry of plants), cell biology, and molecular biology.

Fundamental processes such as photosynthesis, respiration, plant nutrition, plant hormone functions, tropisms, nastic movements, photoperiodism, photomorphogenesis, circadian rhythms, environmental stress physiology, seed germination, dormancy and stomata function and transpiration, both part of plant water relations, are studied by plant physiologists.

Scope

The field of plant physiology includes the study of all the internal activities of plants—those chemical and physical processes associated with life as they occur in plants. This includes study at many levels of scale of size and time. At the smallest scale are molecular interactions of photosynthesis and internal diffusion of water, minerals, and nutrients. At the largest scale are the processes of plant development, seasonality, dormancy, and reproductive control.

Major subdisciplines of plant physiology include phytochemistry (the study of the biochemistry of plants) and phytopathology (the study of disease in plants). The scope of plant physiology as a discipline may be divided into several major areas of research.

First, the study of phytochemistry (plant chemistry) is included within the domain of plant physiology. In order to function and survive, plants produce a wide array of chemical compounds not found in other organisms. Photosynthesis requires a large array of pigments, enzymes, and other compounds to function. Because they cannot move, plants must also defend themselves chemically from herbivores, pathogens and competition from other plants. They do this by producing toxins and foul-tasting or smelling chemicals. Other compounds defend plants against disease, permit survival during drought, and prepare plants for dormancy. While other compounds are used to attract pollinators or herbivores to spread ripe seeds.

Secondly, plant physiology includes the study of biological and chemical processes of individual plant cells. Plant cells have a number of features that distinguish them from cells of animals, and which lead to major differences in the way that plant life behaves and responds differently from animal life. For example, plant cells have a cell wall which restricts the shape of plant cells and thereby limits the flexibility and mobility of plants. Plant cells also contain chlorophyll, a chemical compound that interacts with light in a way that enables plants to manufacture their own nutrients rather than consuming other living things as animals do.

Thirdly, plant physiology deals with interactions between cells, tissues, and organs within a plant. Different cells and tissues are physically and chemically specialized to perform different functions. Roots and rhizoids function to anchor the plant and acquire minerals in the soil. Leaves function to catch light in order to manufacture nutrients. For both of these organs to remain living, the minerals acquired by the roots must be transported to the leaves and the nutrients manufactured in the leaves must be transported to the roots. Plants have developed a number of means by which this transport may occur, such as vascular tissue, and the functioning of the various modes of transport is studied by plant physiologists.

Fourthly, plant physiologists study the ways that plants control or regulate internal functions. Like animals, plants produce chemicals called hormones which are produced in one part of the plant to signal cells in another part of the plant to respond. Many flowering plants bloom at the appropriate time because of light-sensitive compounds that respond to the length of the night, a phenomenon known as photoperiodism. The ripening of fruit and loss of leaves in the winter are controlled in part by the production of the gas ethylene by the plant.

Finally, plant physiology includes the study of how plants respond to conditions and variation in the environment, a field known as environmental physiology. Stress from water loss, changes in air chemistry, or crowding by other plants can lead to changes in the way a plant functions. These changes may be affected by genetic, chemical, and physical factors.

Biochemistry of Plants

The list of simple elements of which plants are primarily constructed—carbon, oxygen, hydrogen, calcium, phosphorus, etc.—is not different from similar lists for animals, fungi, or even bacteria. The fundamental atomic components of plants are the same as for all life; only the details of the way in which they are assembled differs. Despite this underlying similarity, plants produce a vast array of chemical compounds with unusual properties which they use to cope with their environment. Pigments are used by plants to absorb or detect light, and are extracted by humans for use in dyes. Other plant products may be used for the manufacture of commercially important rubber or biofuel. Perhaps the most celebrated compounds from plants are those with pharmacological activity, such as salicylic acid (aspirin), morphine, and digitalis. Drug companies spend billions of dollars each year researching plant compounds for potential medicinal benefits.

Constituent Elements

Plants require some nutrients, such as carbon and nitrogen, in large quantities to survive. Such nutrients are termed macronutrients, where the prefix *macro*-(large) refers to the quantity needed, not the size of the nutrient particles themselves.

Other nutrients, called micronutrients, are required only in trace amounts for plants to remain healthy. Such micronutrients are usually absorbed as ions dissolved in water taken from the soil, though carnivorous plants acquire some of their micronutrients from captured prey.

The following tables list element nutrients essential to plants. Uses within plants are generalized.

Macronutrients. (Necessary in large quantities)

Element	*Form of uptake*	*Notes*
Nitrogen	NO_3^-, NH_4^+	Nucleic acids, proteins, hormones, etc.
Oxygen	O_2 H_2O	Cellulose, starch, other organic compounds
Carbon	CO_2	Cellulose, starch, other organic compounds
Hydrogen	H_2O	Cellulose, starch, other organic compounds
Potassium	K^+	Cofactor in protein synthesis, water balance, etc.
Calcium	Ca^{2+}	Membrane synthesis and stabilization
Magnesium	Mg^{2+}	Element essential for chlorophyll
Phosphorus	$H_2PO_4^-$	Nucleic acids, phospholipids, ATP
Sulfur	SO_4^{2-}	Constituent of proteins and coenzymes

Micronutrients. (Necessary in small quantities)

Element	*Form of uptake*	*Notes*
Chlorine	Cl^-	Photosystem II and stomata function
Iron	Fe^{2+}, Fe^{3+}	Chorophyll formation
Boron	HBO_3	Crosslinking pectin
Manganese	Mn^{2+}	Activity of some enzymes
Zinc	Zn^{2+}	Involved in the synthesis of enzymes and chlorophyll
Copper	Cu^+	Enzymes for lignin synthesis
Molybdenum	MoO_4^{2-}	Nitrogen fixation, reduction of nitrates
Nickel	Ni^{2+}	Enzymatic cofactor in the metabolism of nitrogen compounds

Pigments

Among the most important molecules for plant function are the pigments. Plant pigments include a variety of different kinds of molecules, including porphyrins, carotenoids, and anthocyanins. All biological pigments selectively absorb certain wavelengths of light while reflecting others. The light that is absorbed may be used by the plant to power chemical reactions, while the reflected wavelengths of light determine the colour the pigment will appear to the eye.

Chlorophyll is the primary pigment in plants; it is a porphyrin that absorbs red and blue wavelengths of light while reflecting green. It is the presence and relative abundance of chlorophyll that gives plants their green colour.

All land plants and green algae possess two forms of this pigment: chlorophyll *a* and chlorophyll *b*. Kelps, diatoms, and other photosynthetic heterokonts contain chlorophyll *c* instead of *b*, while red algae possess only chlorophyll *a*.

All chlorophylls serve as the primary means plants use to intercept light in order to fuel photosynthesis.

Carotenoids are red, orange, or yellow tetraterpenoids. They function as accessory pigments in plants, helping to fuel photosynthesis by gathering wavelengths of light not readily absorbed by chlorophyll. The most familiar carotenoids are carotene (an orange pigment found in carrots), lutein (a yellow pigment found in fruits and vegetables), and lycopene (the red pigment responsible for the colour of tomatoes). Carotenoids have been shown to act as antioxidants and to promote healthy eyesight in humans.

Anthocyanins (literally "flower blue") are water-soluble flavonoid pigments that appear red to blue, according to pH. They occur in all tissues of higher plants, providing colour in leaves, stems, roots, flowers, and fruits, though not always in sufficient quantities to be noticeable.

Anthocyanins are most visible in the petals of flowers, where they may make up as much as 30% of the dry weight of the tissue. They are also responsible for the purple colour seen on the underside of tropical shade plants such as *Tradescantia zebrina*; in these plants, the anthocyanin catches light that has passed through the leaf and reflects it back towards regions bearing chlorophyll, in order to maximize the use of available light.

Betalains are red or yellow pigments. Like anthocyanins they are water-soluble, but unlike anthocyanins they are indole-derived compounds synthesized from tyrosine.

This class of pigments is found only in the Caryophyllales (including cactus and amaranth), and never co-occur in plants with anthocyanins. Betalains are responsible for the deep red colour of beets, and are used commercially as food-colouring agents.

Plant physiologists are uncertain of the function that betalains have in plants which possess them, but there is some preliminary evidence that they may have fungicidal properties.

Signals and Regulators

Plants produce hormones and other growth regulators which act to signal a physiological response in their tissues. They also produce compounds such as phytochrome that are sensitive to light and which serve to trigger growth or development in response to environmental signals.

Plant Hormones

Plant hormones, also known as plant growth regulators (PGRs) or phytohormones, are chemicals that regulate a plant's growth. According to a standard animal definition, hormones are signal molecules produced at specific locations, that occur in very low concentrations, and cause altered processes in target cells at other locations. Unlike animals, plants lack specific hormone-producing tissues or organs. Plant hormones are often not transported to other parts of the plant and production is not limited to specific locations.

Plant hormones are chemicals that in small amounts promote and influence the growth, development and differentiation of cells and tissues. Hormones are vital to plant growth; affecting processes in plants from flowering to seed development, dormancy, and germination. They regulate which tissues grow upwards and which grow downwards, leaf formation and stem growth, fruit development and ripening, as well as leaf abscission and even plant death.

The most important plant hormones are abscissic acid (ABA), auxins, gibberellins, and cytokinins, though there are many other substances that serve to regulate plant physiology.

Photomorphogenesis

While most people know that light is important for photosynthesis in plants, few realize that plant sensitivity to light plays a role in the control of plant structural development (morphogenesis). The use of light to control structural development is called photomorphogenesis, and is dependent upon the presence of specialized photoreceptors, which are chemical pigments capable of absorbing specific wavelengths of light.

Plants use four kinds of photoreceptors: phytochrome, cryptochrome, a UV-B photoreceptor, and protochlorophyllide *a*. The first two of these, phytochrome and cryptochrome, are photoreceptor proteins, complex molecular structures formed by joining a protein with a light-sensitive pigment. Cryptochrome is also known as the UV-A photoreceptor, because it absorbs ultraviolet light in the long wave "A" region. The UV-B receptor is one or more compounds that have yet to be identified with certainty, though some evidence suggests carotene or riboflavin as candidates. Protochlorophyllide *a*, as its name suggests, is a chemical precursor of chlorophyll.

The most studied of the photoreceptors in plants is phytochrome. It is sensitive to light in the red and far-red region of the visible spectrum. Many flowering plants use it to regulate the time of flowering based on the length of day and night (photoperiodism) and to set circadian rhythms. It also regulates other responses including the germination of seeds, elongation of seedlings, the size, shape and number of leaves, the synthesis of chlorophyll, and the straightening of the epicotyl or hypocotyl hook of dicot seedlings.

Photoperiodism

Many flowering plants use the pigment phytochrome to sense seasonal changes in day length, which they take as signals to flower. This sensitivity to day length is termed photoperiodism. Broadly speaking, flowering plants can be classified as long day plants, short day plants, or day neutral plants, depending on their particular response to changes in day length. Long day plants require a certain minimum length of daylight to initiate flowering, so these plants flower in the spring or summer. Conversely, short day plants will flower when the length of daylight falls below a certain critical level. Day neutral plants do not initiate flowering based on photoperiodism, though some may use temperature sensitivity (vernalization) instead.

Although a short day plant cannot flower during the long days of summer, it is not actually the period of light exposure that limits flowering. Rather, a short day plant requires a minimal length of uninterrupted darkness in each 24 hour period (a short daylength) before floral development can begin. It has been determined experimentally that a short day plant (long night) will not flower if a flash of phytochrome activating light is used on the plant during the night.

Plants make use of the phytochrome system to sense day length or photoperiod. This fact is utilized by florists and greenhouse gardeners to control and even induce flowering out of season, such as the *Poinsettia*.

Environmental Physiology

Paradoxically, the subdiscipline of environmental physiology is on the one hand a recent field of study in plant ecology and on the other hand one of the oldest. Environmental physiology is the preferred name of the subdiscipline among plant physiologists, but it goes by a number of other names in the applied sciences. It is roughly synonymous with ecophysiology, crop ecology, horticulture and agronomy. The particular name applied to the subdiscipline is specific to the viewpoint and goals of research. Whatever name is applied, it deals with the ways in which plants respond to their environment and so overlaps with the field of ecology.

Environmental physiologists examine plant response to physical factors such as radiation (including light and ultraviolet radiation), temperature, fire, and wind. Of particular importance are water relations (which can be measured with the Pressure bomb) and the stress of drought or inundation, exchange of gases with the atmosphere, as well as the cycling of nutrients such as nitrogen and carbon.

Environmental physiologists also examine plant response to biological factors. This includes not only negative interactions, such as competition, herbivory, disease and parasitism, but also positive interactions, such as mutualism and pollination.

Tropisms and Nastic Movements

Plants may respond both to directional and nondirectional stimuli. A response to a directional stimulus, such as gravity or sunlight, is called a tropism. A response to a nondirectional stimulus, such as temperature or humidity, is a nastic movement.

Tropisms in plants are the result of differential cell growth, in which the cells on one side of the plant elongate more than those on the other side, causing the part to bend toward the side with less growth. Among the common tropisms seen in plants is phototropism, the bending of the plant toward a source of light. Phototropism allows the plant to maximize light exposure in plants which require additional light for photosynthesis, or to minimize it in plants subjected to intense light and heat. Geotropism allows the roots of a plant to determine the direction of gravity and grow downwards. Tropisms generally result from an interaction between the environment and production of one or more plant hormones.

In contrast to tropisms, nastic movements result from changes in turgor pressure within plant tissues, and may occur rapidly. A familiar example is thigmonasty (response to touch) in the Venus fly trap, a carnivorous plant. The traps consist of modified leaf blades which bear sensitive trigger hairs. When the hairs are touched by an insect or other animal, the leaf folds shut. This mechanism allows the plant to trap and digest small insects for additional nutrients. Although the trap is rapidly shut by changes in internal cell pressures, the leaf must grow slowly in order to reset for a second opportunity to trap insects.

Plant Disease

Economically, one of the most important areas of research in environmental physiology is that of phytopathology, the study of diseases in plants and the manner in which plants resist or cope with infection. Plant are susceptible to the same kinds of disease organisms as animals, including viruses, bacteria, and fungi, as well as physical invasion by insects and roundworms. Because the biology of plants differs from animals, their symptoms and responses are quite different. In some cases, a plant can simply shed infected leaves or flowers to prevent to spread of disease, in a process called abscission. Most animals do not have this option as a means of controlling disease. Plant diseases organisms themselves also differ from those causing disease in animals because plants cannot usually spread infection through casual physical contact. Plant pathogens tend to spread via spores or are carried by animal vectors.

One of the most important advances in the control of plant disease was the discovery of Bordeaux mixture in the nineteenth century. The mixture is the first known fungicide and is a combination of copper sulfate and lime. Application of the mixture served to inhibit the growth of downy mildew that threatened to seriously damage the French wine industry.

History

Early History

Sir Francis Bacon published one of the first plant physiology experiments in 1627 in the book, *Sylva Sylvarum.* Bacon grew several terrestrial plants, including a rose, in water and concluded that soil was only needed to keep the plant upright. Jan Baptist van Helmont published what is considered the first quantitative experiment in plant physiology in 1648. He grew a willow tree for five years in a pot containing 200 pounds of oven-dry soil.

The soil lost just two ounces of dry weight and van Helmont concluded that plants get all their weight from water, not soil. In 1699, John Woodward published experiments on growth of spearmint in different sources of water. He found that plants grew much better in water with soil added than in distilled water. Stephen Hales is considered the Father of Plant Physiology for the many experiments in the 1727 book; though Julius von Sachs unified the pieces of plant physiology and put them together as a discipline. His *Lehrbuch der Botanik* was the plant physiology bible of its time.

Researchers discovered in the 1800s that plants absorb essential mineral nutrients as inorganic ions in water. In natural conditions, soil acts as a mineral nutrient reservoir but the soil itself is not essential to plant growth. When the mineral nutrients in the soil are dissolved in water, plant roots absorb nutrients readily, soil is no longer required for the plant to thrive. This observation is the basis for hydroponics, the growing of plants in a water solution rather than soil, which has become a standard technique in biological research, teaching lab exercises, crop production and as a hobby.

Current Research

One of the leading journals in the field is *Plant Physiology*, started in 1926. All its back issues are available online for free. Many other journals often carry plant physiology articles, including *Physiologia Plantarum, Journal of Experimental Botany, American Journal of Botany, Annals of Botany, Journal of Plant Nutrition* and *Proceedings of the National Academy of Sciences.*

Economic Applications

Food Production

In horticulture and agriculture along with food science, plant physiology is an important topic relating to fruits, vegetables, and other consumable parts of plants. Topics studied include: *climatic* requirements, fruit drop, nutrition, ripening, fruit set. The production of food crops also hinges on the study of plant physiology covering such topics as Optimal planting and harvesting times and post harvest storage of plant products for human consumption and the production of secondary products like drugs and cosmetics.

Plant Nutrition

Plant nutrition is the study of the chemical elements that are necessary for plant growth. There are several principles that apply to plant nutrition. Some elements are directly involved in plant metabolism. However, this principle does not account for the so-called beneficial elements, whose presence, while not required, has clear positive effects on plant growth.

A nutrient that is able to limit plant growth according to Liebig's law of the minimum, is considered an essential plant nutrient if the plant can not complete its full life cycle without it. There are 14 essential plant nutrients.

Carbon, Hydrogen and Oxygen are essential for plants, but are not considered plant nutrients.

Macronutrients:

- N = Nitrogen
- K = Potassium
- Ca = Calcium
- P = Phosphorus
- Mg = Magnesium
- S = Sulfur
- Si = Silicon

Micronutrients (trace levels) include:

- Cl = Chlorine
- Fe = Iron
- B = Boron
- Mn = Manganese
- Na = Sodium
- Zn = Zinc
- Cu = Copper
- Ni= Nickel
- Mo = Molybdenum.

These nutrients are further divided into the mobile and immobile nutrients. A plant will always supply more nutrients to its younger leaves than its older ones, so when nutrients are mobile, the lack of nutrients is first visible on older leaves. When a nutrient is less mobile, the younger leaves suffer because the nutrient does not move up to them but stays lower in the older leaves. Nitrogen, phosphorus, and potassium are mobile nutrients, while the others have varying degrees of mobility.

Processes

Plants uptake essential elements from the soil through their roots and from the air through their leaves. Nutrient uptake in the soil is achieved by cation exchange, wherein root hairs pump hydrogen ions (H^+) into the soil through proton pumps. These hydrogen ions displace cations attached to negatively charged soil particles so that the cations are available for uptake by the root. In the leaves, stomata open to take in carbon dioxide and expel oxygen. The carbon dioxide molecules are used as the carbon source in photosynthesis.

Though nitrogen is plentiful in the earth's atmosphere, relatively few plants engage in nitrogen fixation (conversion of atmospheric nitrogen to a biologically useful form). Most plants therefore require nitrogen compounds to be present in the soil in which they grow.

Functions of Nutrients

Each of these nutrients is used in a different place for a different essential function.

Carbon: Carbon forms the backbone of many plants biomolecules, including starches and cellulose. Carbon is fixed through photosynthesis from the carbon dioxide in the air and is a part of the carbohydrates that store energy in the plant.

Hydrogen: Hydrogen also is necessary for building sugars and building the plant. It is obtained almost entirely from water.

Oxygen: Oxygen is necessary for cellular respiration. Cellular respiration is the process of generating energy-rich adenosine triphosphate (ATP) via the consumption of sugars made in photosynthesis. Oxygen gas is produced as a by-product from this reaction.

Phosphorus: Phosphorus is important in plant bioenergetics. As a component of ATP, phosphorus is needed for the conversion of light energy to chemical energy (ATP) during photosynthesis. Phosphorus can also be used to modify the activity of various enzymes by phosphorylation, and can be used for cell signalling. Since ATP can be used for the biosynthesis of many plant biomolecules, phosphorus is important for plant growth and flower/seed formation.

Potassium: Potassium regulates the opening and closing of the stoma by a potassium ion pump. Since stomata are important in water regulation, potassium reduces water loss from the leaves and increases drought tolerance. Potassium deficiency may cause necrosis or interveinal chlorosis.

Nitrogen: Nitrogen is an essential component of all proteins. Nitrogen deficiency most often results in stunted growth.

Sulphur: Sulphur is a structural component of some amino acids and vitamins, and is essential in the manufacturing of chloroplasts.

Silicon: Silicon is deposited in cell walls and contributes to its mechanical properties including rigidity and elasticity

Calcium: Calcium regulates transport of other nutrients into the plant and is also involved in the activation of certain plant enzymes. Calcium deficiency results in stunting.

Magnesium: Magnesium is an important part of chlorophyll, a critical plant pigment important in photosynthesis. It is important in the production of ATP through its role as an enzyme cofactor. There are many other biological roles for magnesium. Magnesium deficiency can result in interveinal chlorosis.

Iron: Iron is necessary for photosynthesis and is present as an enzyme cofactor in plants. Iron deficiency can result in interveinal chlorosis and necrosis.

Molybdenum: Molybdenum is a cofactor to enzymes important in building amino acids.

Boron: Boron is important in sugar transport, cell division, and synthesizing certain enzymes. Boron deficiency causes necrosis in young leaves and stunting.

Copper: Copper is important for photosynthesis. Symptoms for copper deficiency include chlorosis. Involved in many enzyme processes. Necessary for proper photosythesis. Involved in the manufacture of lignin (cell walls). Involved in grain production.

Manganese: Manganese is necessary for building the chloroplasts. Manganese deficiency may result in coloration abnormalities, such as discoloured spots on the foliage.

Sodium: Sodium is involved in the regeneration of phosphoenolpyruvate in CAM and C4 plants. It can also substitute for potassium in some circumstances.

Zinc: Zinc is required in a large number of enzymes and plays an essential role in DNA transcription. A typical symptom of zinc deficiency is the stunted growth of leaves, commonly known as "little leaf" and is caused by the oxidative degradation of the growth hormone auxin.

Nickel: In higher plants, Nickel is essential for activation of urease, an enzyme involved with nitrogen metabolism that is required to process urea. Without Nickel, toxic levels of urea accumulate, leading to the formation of necrotic lesions. In lower plants, Nickel activates several enzymes involved in a variety of processes, and can substitute for Zinc and Iron as a cofactor in some enzymes.

Cobalt has proven to be beneficial to at least some plants, but is essential in others, such as legumes where it is required for nitrogen fixation. Vanadium may be required by some plants, but at very low concentrations. It may also be substituting for molybdenum. Selenium and sodium may also be beneficial. Sodium can replace potassium's regulation of stomatal opening and closing.

Plant nutrition is a difficult subject to understand completely, partially because of the variation between different plants and even between different species or individuals of a given clone. Elements present at low levels may cause deficiency symptoms, and toxicity is possible at levels that are too high.

Further, deficiency of one element may present as symptoms of toxicity from another element, and vice-versa. Carbon and oxygen are absorbed from the air, while other nutrients are absorbed from the soil. Green plants obtain their carbohydrate supply from the carbon dioxide in the air by the process of photosynthesis.

Chapter 6

Micropropagation in Plants

Micropropagation is the practice of rapidly multiplying stock plant material to produce a large number of progeny plants, using modern plant tissue culture methods.

Micropropagation is used to multiply novel plants, such as those that have been genetically modified or bred through conventional plant breeding methods. It is also used to provide a sufficient number of plantlets for planting from a stock plant which does not produce seeds, or does not respond well to vegetative reproduction.

Methods

Establishment

Micropropagation begins with the selection of plant material to be propagated. Clean stock materials that are free of viruses and fungi are important in the production of the healthiest plants. Once the plant material is chosen for culture, the collection of explant(s) begins and is dependent on the type of tissue to be used; including stem tips, anthers, petals, pollen and others plant tissues. The explant material is then surface sterilized, usually in multiple courses of bleach and alcohol washes and finally rinsed in sterilized water. This small portion of plant tissue, sometimes only a single cell, is placed on a growth medium, typically containing sucrose as an energy source and one or more plant growth regulators (plant hormones). Usually the medium is thickened with agar to create a gel which supports the explant during growth. Some plants are easily grown on simple media but others require more complicated media for successful growth; The plant tissue grows and differentiates into new tissues depending on the medium. For example, media containing cytokinins are used to create branched shoots from plant buds and it happens in a vegetative form.

Multiplication

Multiplication is the taking of tissue samples produced during the first stage and increasing their number. Following the successful introduction and growth of plant tissue, the establishment stage is followed by multiplication. Through repeated cycles of this process, a single explant sample may be increased from one to hundreds or thousands of plants. Depending on the type of tissue grown, multiplication can involve different methods and media. If the plant material grown is callus tissue, it can be placed in a blender and cut into smaller pieces and recultured on the same type of culture medium to grow more callus tissue. If the tissue is grown as small plants called plantlets, hormones are often added that cause the plantlets to produce many small offshoots that can be removed and recultured.

Pretransplant

This stage involves treating the plantlets/shoots produced to encourage root growth and "hardening." It is performed *in vitro*, or in a sterile "test tube" environment."Hardening" refers to the preparation of the plants for a natural growth environment. Until this stage, the plantlets have been grown in "ideal" conditions, designed to encourage rapid growth.

Due to lack of necessity, the plants are likely to be highly susceptible to disease and often do not have fully functional dermal coverings and will be inefficient in their use of water and energy. In vitro conditions are high in humidity and plants grown under these condition do not form a working cuticle and stomata that keep the plant from drying out, when taken out of culture the plantlets need time to adjust to more natural environmental conditions. Hardening typically involves slowly weaning the plantlets from a high-humidity, low light, warm environment to what would be considered a normal growth environment for the species in question. This is done by moving the plants to a location high in humidity,

Transfer from Culture

In the final stage of plant micropropagation, the plantlets are removed from the plant media and transferred to soil or (more commonly) potting compost for continued growth by conventional methods.

This stage is often combined with the "pretransplant" stage.

Advantages

Micropropagation has a number of advantages over traditional plant propagation techniques:

- The main advantage of micropropagation is the production of many plants that are clones of each other.
- Micropropagation can be used to produce disease-free plants.
- Micropropagation produces rooted plantlets ready for growth, saving time for the grower when seeds or cuttings are slow to establish or grow.
- It can have an extraordinarily high fecundity rate, producing thousands of propagules while conventional techniques might only produce a fraction of this a number.
- It is the only viable method of regenerating genetically modified cells or cells after protoplast fusion.
- It is useful in multiplying plants which produce seeds in uneconomical amounts, or when plants are sterile and do not produce viable seeds or when seed can't be stored (vgr. recalcitrant seeds).
- Micropropagation often produces more robust plants, leading to accelerated growth compared to similar plants produced by conventional methods - like seeds or cuttings.
- Some plants with very small seeds, including most orchids, are most reliably grown from seed in sterile culture.
- A greater number of plants can be produced per square meter and the propagules can be stored longer and in a smaller area.

Disadvantages

Micropropagation is not always the perfect means of multiplying plants, conditions that limits its use include:

- It is *very* expensive, and can have a labor cost of more than 70%
- A monoculture is produced after micropropagation, leading to a lack of overall disease resilience, as all progeny plants may be vulnerable to the same infections.
- An infected plant sample can produce infected progeny. This is uncommon if the stock plants are carefully screened and vetted to prevent culturing plants infected with virus or fungus.
- Not all plants can be successfully tissue cultured, often because the proper medium for growth is not known or the plants produce secondary metabolic chemicals that stunt or kill the explant.

- Sometimes plants or cultivars do not come true to type after being tissue cultured, this is often dependent on the type of explant material utilized during the initiation phase or the result of the age of the cell or propagule line.
- Some plants are very difficult to disinfest of fungal organisms.

The major limitation in the use of Micropropagation for many plants is the cost of production; for many plants the use of seeds, which are normally disease free and produced in good numbers, readily produce plants in good numbers at a lower cost. For this reason, many plant breeders do not utilize micropropagation because the cost is prohibitive, other breeders use it to produce stock plants that are then used for seed multiplication. Mechanisation of the process could reduce labour costs, but has proven difficult to achieve, despite active attempts to develop technological solutions.

Plant Propagation

Plant propagation is the process of creating new plants from a variety of sources: seeds, cuttings, bulbs and other plant parts. Plant propagation can also refer to the artificial or natural dispersal of plants.

Sexual Propagation (Seed)

Seeds and spores can be used for reproduction (through e.g. sowing). Seeds are typically produced from sexual reproduction within a species, because genetic recombination has occurred plants grown from seeds may have different characteristics from its parents. Some species produce seeds that require special conditions to germinate, such as cold treatment. The seeds of many Australian plants and plants from southern Africa and the American west require smoke or fire to germinate. Some plant species, including many trees do not produce seeds until they reach maturity, which may take many years. Seeds can be difficult to acquire and some plants do not produce seed at all.

Electrical Propagation

Commonly used both by hobby gardeners and in large scale industrial facilities, an electrical propagator is a low-cost, efficient and safe method of encouraging seed growth.

Asexual Propagation

Plants have a number of mechanisms for asexual or vegetative reproduction. Some of these have been taken advantage of by horticulturists and gardeners to multiply or clone plants rapidly. People also use methods that plants do not use, such as tissue culture and grafting. Plants are produced using material from a single parent and as such there is no exchange of genetic material, therefore vegetative propagation methods almost always produce plants that are identical to the parent. Vegetative reproduction uses plants parts such as roots, stems and leaves. In some plants seeds can be produced without fertilization and the seeds contain only the genetic material of the parent plant. Therefore, propagation via asexual seeds or apomixis is asexual reproduction but not vegetative propagation.

Techniques for vegetative propagation include:

- Air or ground layering
- Division
- Grafting and bud grafting, widely used in fruit tree propagation
- Micropropagation
- Stolons or runners
- Storage organs such as bulbs, corms, tubers and rhizomes
- Striking or cuttings
- Twin-scaling.

Heated Propagator

A heated propagator is a horticultural device to maintain a warm and damp environment for seeds and cuttings to grow in. This can be in the form of a clear enclosed bin sitting over a hotpad, or even a portable heater pointed at the bin. the key is to keep the moisture in the clear bin, while keeping lighting over the top of it, usually.

Seed Propagation Mat

An electric seed-propagation mat is a heated rubber mat covered by a metal cage which is used in gardening. The mats are made so that planters containing seedlings can be placed on top of the metal cage without the risk of starting a fire. In extreme cold, gardeners place a loose plastic cover over the planters/mats which creates a sort of miniature greenhouse. The constant and predictable heat allows people to garden in the winter months when the weather is generally too cold for seedlings to survive naturally. When combined with a lighting system, many plants can be grown indoors using these mats.

Tissue Culture of Woody Plants

Much has been written and more has been said about the applications of plant tissue culture to the nursery industry. The subject is still poorly understood by a majority of plant propagators. Tissue culture is often looked upon as being practical only for propagation of those plants which are difficult to propagate by conventional methods. This notion is quickly being laid to rest.

An industry is emerging in various areas throughout the country which is taking advantage of technological advances in the realm of tissue culture of woody plants. This technology suggests that propagation by tissue culture is indeed applicable to the "difficult to propagate" species, and equally as important, it may offer economic advantages for some species which are considered relatively "easy to propagate".

The intent of this communication is to briefly describe the contemporary techniques of tissue culture of woody plants, and to review the advantages and limitations of tissue culture as a propagation method.

The Methodology of Tissue Culture

The variety of techniques which can be used to get plant development in vitro (that is, by techniques such as tissue culture) is considerable and completely dependent upon the species in question. Single cells of leaf tissue can regenerate whole plants, as can shoot tips, leaf pieces, root pieces, lateral buds, or stem sections. Not all of these methods are applicable to woody plants, and indeed, not all have been applied to a commercial level to any plants.

Shoot tip culture is the method in widest use for the mass propagation of woody species. An actively growing shoot tip is surface sterilized and placed on a defined culture medium under sterile conditions. The culture medium contains inorganic and organic salts (macronutrients, micronutrients and vitamins) as well as an energy source, growth regulators, and agar)to gel the medium). If the growth regulators are appropriately balanced, the shoot tip elongates, lateral buds break and begin growth, and adventitious shoots are also produced on the stem piece. This rapid proliferation of shoots results in masses of shoots being produced from a single shoot tip. Up to a hundred shoots may be produced in as little as eight to twelve weeks from a single tip. The number of shoots produced and the rapidity of shoot proliferation varies between species, and in some cases, between cultivars of a single species. Shoots are removed from the cultures at regular intervals and a portion of the mass is replaced on fresh media to continue proliferation. The small shoots which are removed are then rooted in a separate medium, either a sterile gelled medium or a peat-perlite medium (in much the same fashion as conventional woody cuttings).

Shoots produced through tissue culture are generally easy to root, even though the same cultivar may be difficult to root by cutting propagation. The resulting rooted shoot is referred to as a "plantlet" because of its miniature size, but these can generally be grown in the greenhouse at a rapid rate and with a high degree of survival. Techniques for rooting tissue cultured shoots are currently receiving a great deal of research attention as are methods for establishing these shoots into the greenhouse environment.

The Tissue Culture Laboratory

Most propagators would probably list the expense of setting up a tissue culture laboratory among the reasons for not getting involved with this technology. A lab does cost more than a mist bed! A more severe limitation is without doubt the absence of personnel skilled in the techniques of tissue culture. An individual who has had prior experience with tissue culture methodology can successfully operate a laboratory with a bare minimum of equipment. An autoclave or sterilizer, an accurate means of weighing chemicals, a small inventory of glassware, and chemical stocks or prepackaged media are all that is required to initiate cultures. The laboratory can be enlarged, more equipment added, and additional personnel hired as the need for such arises. The most critical factors are the ingenuity and experience of the laboratory manager.

As a tissue culture laboratory becomes operational, procedures other than the actual culturing process become critical. Maintenance of disease and insect free stock plants with a low titter of bacterial and fungal contamination is critical. Also important is the care and handling of the thousands of plantlets prior to their transfer to the field or containers. The ability to grow off the plantlets ultimately will limit production, rather than any limitation due to the propagation phase.

Tissue Culture Propagation of Woody Species

Major advances in the mass propagation of woody plants have been made over the past ten years. Because of the somewhat variable response of species and cultivars to the in vitro environment, much research was (and still is) needed to define the cultural conditions required by ornamental species. As an example of the variability encountered, we have observed that seedling shoot tip cultures of crabapple produce as few as one or as many as sixty-four shoots per explant after three months in tissue culture. Similar variations in the tendency for multiple shoot production have also been noted with adult clonal material of different apple cultivars. The number of woody plant species which have been clonally propagated through tissue culture is increasing at a rapid rate, and in fact, most commercially important ornamental species have been studied. Perhaps the best publicized of the early research on woody plants involved apples, Douglas fir, and rhododendrons. An extensive commercial scale production program for apple scion cultivars and rootstocks has developed in England and in several European countries.

Using the successes of the apple, rhododendron, and maple programs as a basis, many recent advances have been made which will facilitate the mass propagation of other woody species. While it is true that cultivar differences account for lack of success with some plants, a growing body of evidence suggests that taxonomic families behave somewhat alike in tissue culture.

A large number of deciduous and evergreen Rhododendron species share common cultural requirements, the most notable of which is the requirement for the same cytokinin for adventitious shoot formation. A similar relationship exists between the trees and shrubs of the Rosaceae (including crabapples, pears, plums, and hawthorns). Most rosaceous species will likely be mass propagated on media very similar to that found to be optimal for apples.

The anomalies alluded to by the crabapple seedling responses to tissue culture and the emerging similarities between the requirements of rosaceous trees and shrubs demonstrate that much research is needed to define the conditions necessary for commercial scale tissue culture production programs. As is often found to be the case with propagation systems, those systems described as optimal in the laboratory would likely need modification to be considered economical by the industry.

Impact of the Tissue Culture Industry

As is always the case for any emerging technology, the strongest supporters for the adoption of that technology are likely those involved in its development. Having acknowledged these biases, one must now consider the advantages and limitations of tissue culture propagation, and what its economic impact on the nursery industry will be.

Not every nursery will Operate a tissue culture laboratory. The trend toward specialization of nursery operations as either "propagators" or "producers" is likely to continue in the future. Operation and management of a tissue culture facility requires personnel and physical facilities which are quite different from those required by a container or field production nursery. Previously in this report where the establishment of the laboratory was considered, the relatively small initial investment in physical facilities and equipment was discussed. It should be reemphasized at this point, that the training and quality of the laboratory manager is more important than the laboratory itself, and that this expertise is not currently available among contemporary nursery propagators.

Not every nursery will Need a tissue culture laboratory. Commercial operations specializing in mass propagation of nursery crops can likely produce a sufficient number of plants to satisfy the market.

As with any other technology, the success of tissue culture within the nursery industry will be determined by economic factors. Can the plantlets be produced in sufficient numbers as a cost to compare with traditional propagation? Will the propagules fit into the established production sequence in the nursery industry? Are the propagules true-to-type? These answers must be obtained on a species by species, cultivar by cultivar basis.

Care and Handling of Micropropagated Plants

The propagation of plants through tissue culture, micropropagation, was viewed at one time as a research tool to study plant growth and development, growth regulators, and organogenesis. Much useful information was gathered as a result of such research. The impact which micropropagation had on the multiplication of orchids, tropical foliage plants, and selected floricultural crops is now being felt in the nursery crops industry. Rhododendrons, azaleas, mountain laurel, lilacs, and a few deciduous trees are currently being micropropagated. Tissue cultured plants are now accepted without fear of extreme variation, poor vigor, or delay in flowering as was the prior assumption.

Advantages of Micropropagated Plants

Certain morphological and physiological characteristics of tissue cultured plants make them superior to conventionally propagated materials. Tissue cultured plants (particularly Amelanchier, Rhododendron, and Betula) tend toward a more branched habit. This tendency results in a young plant which is less "leggy" and salable earlier because of increased fullness. Tissue cultured plants produce cuttings which have increased rooting when rooted by conventional means. This phenomenon is being exploited in the "rejuvenation" of stock plants for cutting propagation. The most obvious advantage inherent to micropropagation is that more plants are produced quicker from a limited number of stock plants (one in some cases). New introductions are available in commercial quantities in a fraction of the time previously required.

The Process of Micropropagation

A brief review of the procedure of tissue culture will clarify the need for specific care and handling. A small piece of the plant to be cloned (the explant) is removed from a healthy, well-maintained stock plant and sterilized in a dilute bleach solution. The source of tissue will vary by species, but shoot tip, leaf, stem, lateral bud, and flower tissues have been used successfully for various plants. Rhododendron, for example, will produce plantlets from flower petal tissue. The sterilized explant is rinsed with sterile water, and placed in aseptically prepared containers on a specially formulated medium. The explant may produce shoot proliferating cultures directly by enhanced lateral bud break, or the tissue may undergo a certain period of unorganized growth (callus) prior to shoot differentiation. The pattern of growth of the cultures is principally determined by the plant growth regulator content of the tissue culture medium (the auxin and cytokinin concentration). Most cultures are established within 4 to 12 weeks depending on the species and in some instances, depending on the cultivar. A shoot proliferating culture is one which can be subdivided (subcultured) to produce divisions which will continue rapid multiplication. Estimates of the rate of multiplication vary and are affected by many factors, but the production of thousands and in some cases millions of plants a year from a single explant has been demonstrated!

Morphology of A Microcutting

The small shoots produced in tissue culture are referred to as "microcuttings". Microcuttings vary a great deal in size, length and degree of leaf expansion depending upon species, cultivar, and cultural conditions. The tissue culture environment results in the production of shoots at a continuum of stages of development, from ones just recently expanded to those with elongated stems and well expanded leaves. Recognition of the type of microcutting which roots and grows out well is one of the most critical components of a successful system. The determination of the correct sequence of media and/or transfers to optimize the production of "normal" microcuttings that root and acclimate in high percentage is important. Thousands of small, spindly microcuttings which root and grow out poorly are worthless in the commercial propagation arena.

Choice of a Rooting Method

Microcuttings are rooted by modifications of two basic methods. In vitro rooting is accomplished by transferring the carefully dissected microcuttings to a medium which is free of growth regulators or has only auxin added as the growth regulator.

All manipulations are conducted under sterile conditions, the process requires excessive labor input, and media must be prepared specifically for the rooting stage. Since the cuttings are provided with minerals, vitamins, sugar as a carbon source, and root promoting auxins, many researchers feel this method will allow for rooting of even the most difficult to root plants. This method does allow more flexibility in choosing cuttings since cuttings need not conduct photosynthesis to provide an energy source for rooting. Leaf expansion is therefore less critical.

In most commercial operations, the preferred rooting method is referred to as nonsterile rooting. Microcuttings are harvested (usually under nonsterile conditions by sacrificing whole cultures, but if stocks are low, sterility may be maintained at this stage) and stuck in a clean (but not sterile) mixture of peat moss and perlite or vermiculite, or in preformed foam or peat moss plugs. Cuttings may be stuck in small plastic covered containers and rooted in the lab, or placed directly into flats which are then placed in a mist, fog, or poly tent system. The nonsterile method has given outstanding results with most species tested. Rooting is accomplished in 3 to 5 weeks, and the microcuttings can be transplanted at this stage and then acclimated or acclimated directly in the rooting flats.

Acclimation of Rooted Microcuttings

Microcuttings are very sensitive to changes in the physical environment, and the success or failure of the whole operation may hinge on the acclimation process. Microcuttings are poorly adapted for growth in the greenhouse environment at the time they are removed from the rooting containers, whether these were held in the laboratory or in the mist or fog area. Microcuttings rooted and grown in very high humidity have stomates which have been shown to be "sluggish", responding to decreased relative humidity too slowly to prevent desiccation of the cutting. This problem is compounded by the apparent lack of a well developed, waxy cuticle, the barrier to water loss over the entire plant surface. Research has shown that the cuticle is present, but not functional to the extent that exists on seedling or even cutting propagated material.

Microcuttings must be acclimated to increased light intensity in much the same manner as acclimation to decreased relative humidity. If too much leaf surface is injured by direct transferal to higher light intensity (sun-burning), the vigor of the cutting will be reduced markedly because the plant is too small to have enough stored starch to force a new flush of growth.

The following protocol has been successful for a number of woody species. Rooted microcuttings are transplanted to plug trays, sectioned flats, or peat moss pots into a commercially prepared medium (usually a peat moss and perlite or peat-vermiculite medium; it is important that the medium be fine textured, similar to a seed starting mix). This operation must be accomplished quickly, and care must be taken to fog or mist the microcuttings to "water them in". Transplanted flats are carried to the mist bed just as soon as they are filled. Microcuttings which are allowed to wilt during the transplanting process show poor recovery, and poor transplant survival!

The mist cycle is set for a duration of 4 or 6 seconds with the interval set at 4 or 6 minutes. Bottom heat is beneficial. The mist area must be shaded with heavy shade, with a light exclusion of at least 60%. Microcuttings are left in the mist + shade for 5 to 7 days (some species may require 2 or 3 more days if there is any sign of wilting upon removal) and they are transferred to a shade house (shade greater than 50% light exclusion) for an additional 7 days.

Microcuttings usually begin a flush of growth near the end of the time in the shade house. Modifications of this procedure can be made for fog or poly tent rooted microcuttings, but the transition from high to low relative humidity and low to high light must be gradual. Microcuttings which are forced into a growth cessation by improper acclimation procedures have low survival percentages, and tend to grow out slowly and nonuniformly.

Seasonal Fluctuations in Acclimation Efficiency

Micropropagation is unique among propagation methods in that seasonality has little influence on the production of microcuttings in the lab. Most woody species, however, do display seasonal differences in acclimation efficiency and subsequent growth in the greenhouse. Microcuttings acclimated between February and September establish well in the greenhouse and generally grow out rapidly and uniformly. Microcuttings acclimated between October and January acclimate adequately, but usually do not put on the first growth flush uniformly, and generally go into a semidormant condition which requires a cold treatment to break. Several techniques have been used in attempting to break this cycle. HPS lighting to supplement natural daylengths to 18 hours of light per day, constant feed fertilization, and incandescent light breaks have all failed to coax lagging microcuttings into active growth. We have concluded that light intensity, rather than absolute daylength, must be governing the early growth response of the acclimated microcuttings.

The current recommendation is to schedule microcuttings to grow out in the greenhouse between February and September to avoid nonuniform growth following acclimation.

Should Production Nurseries Purchase Shoot Proliferating Cultures?

The concern over whether production nurseries should purchase proliferating cultures directly from tissue culture labs has arisen on many occasions recently. Shoot proliferating cultures are a convenient, compact means of transporting cuttings, they are "clean", and are available to a much lower unit cost since the cost of the rooting and acclimation stages are not included.

The current recommendation is that production nurseries unfamiliar with rooting of tissue cultured materials not attempt rooting and acclimation of microcuttings without first conducting limited trials to gain experience in the procedure.

Consultation with individuals who have developed rooting and acclimation techniques has revealed that the first few experiments usually result in less than acceptable results. Individuals not familiar with the sensitivity of microcuttings to shifts in relative humidity, light intensity, and the importance of cutting selection generally will not experience the level of success necessary to make the process commercially viable. Once the tissue cultured plants have formed the first growth flush (rooted liner), they could be handled as any transplant and would not require any increased skill level from the grower.

Micropropagation is rapidly becoming a standard method in the nursery crops industry, much as it has in the foliage plants industry. The success or failure of the technique at the commercial level is often not in the lab, but rather in the greenhouse at the level of rooting and acclimation of microcuttings. Microcuttings are very sensitive to changes in environmental conditions in the early stages, but respond just as any other propagule after the first growth flush. Production nurseries should purchase tissue cultured plants at the rooted liner stage, unless a suitable system has been developed to root and acclimate microcuttings from the shoot proliferating culture stage.

Field Transplant Survival of *Amelanchier* Liners Produced by Tissue Culture

Propagation of nursery crops through tissue culture is gaining increasing acceptance from the nursery industry. Micropropagation systems from initial explanting to acclimation of microcuttings have been developed for many species.

Physiological and morphological changes which occur in micropropagated plants during acclimation from aseptic conditions to the greenhouse environment have focused on two areas. The structure and amount of epicuticular was on the leaves increases upon exposure of microcuttings to lowered relative humidity, a change deemed necessary to lower rates of water loss from the delicate microcuttings. Concomitantly, the stomata increase in rate of response to water stress after 5 days of acclimation. This change is believed to be significant to the survival of tissue cultured woody plants.

Relatively little research has been conducted to determine factors necessary to acclimation of tissue cultured liners to the field environment. Smith recommends a minimum size of 10-15 cm top growth and an 8-10 cm root ball for field transplanting of tissue cultured liners. Dunstan reported that tissue cultured apple rootstocks transplanted to the field directly from the greenhouse in early spring were susceptible to frost damage and were defoliated but regrew after about 2 months of quiescence. Dunstan's transplanting procedure for summer transplanting now involves preconditioning plants under 50% shade to achieve a "partially exposed environment into which plants are placed before full exposure". Losses of tissue cultured liners were no greater than those for seedlings using this procedure.

Relatively little research has been conducted to determine factors necessary to acclimation of tissue cultured liners to the field environment. Smith recommends a minimum size of 10-15 cm top growth and an 9-10 cm root ball for field transplanting of tissue cultured liners. Dunstan reported that tissue cultured apple rootstocks transplanted to the field directly from the greenhouse in early spring were susceptible to frost damage and were defoliated but regrew after about 2 months of quiescence.

Dunstan's transplanting procedure for summer transplanting now involves preconditioning plants under 50% shade to achieve a "partially exposed environment into which plants are placed before full exposure". Losses of tissue cultured liners were no greater than those for seedlings using this procedure.

This research was undertaken to determine the effect of certain factors (container type, planting date, days of outdoor acclimation) on the transplant survival of tissue cultured plants of *Amelanchier laevis.*

Materials and Methods

Microcuttings of *Amelanchier laevis* were produced in vitro according to the method described for Hally Jolivette cherry. The growth regulator concentration in the shoot proliferation stage was 0.1 mg/1 naphthaleneacetic acid and 2.5 mg/1 benzladenine.

These microcuttings were rooted in plastic-covered foil containers containing a moistened peatmoss-perlite medium (1:1 by volume). Rooting was done in the tissue culture laboratory where the temperature was 24-27¡C and lighting was provided for 16 hr per day at 40μ Einsteins/M2/sec by cool white fluorescent lamps.

Transplanted microcuttings were immediately placed on a shaded intermittent mist bench for 3 days (6 sec most every 6 min; light intensity, 360μ Einsteins/M2/sec). After an additional 7 to 10 days of acclimation on a shaded greenhouse bench (light intensity, 270μ Einsteins/m2/sec), plants were grown under standard greenhouse cultural conditions including fertilization at every watering with 200 ppm of 20-20-20 soluble fertilizer. Plants which received an outdoor acclimation treatment were held for various time intervals under 60% shade (light intensity, 810μ Einsteins/m2/sec) prior to planting.

Field transplanting studies were conducted on the Ohio State University Dept. of Horticulture's Lane Avenue Farm. The plot used was a Brookston silty clay loam with a previous record of heavy fertilization. The plot was irrigated as needed.

In experiment 1, plantlets approximately 15 cm tall were acclimated for 0, 5, 10, or 15 days under 60% shade outdoors. Two types of containers were used, 5.7 x 5.7 cm peatmoss pots or 4 x 20 cm Cone-Tainers (Ray Leach Cone-Tainer Nursery, Canby, OR 97013). A 1:1 (by volume) peatmoss-perlite medium was used. Field planting was done June 28, 1981.

A randomized complete block design with six six-plant replications was used. Experiment 2 was a replication of experiment 1 except only peatmoss pots were used. Field planting was done August 6, 1981. A randomized complete block design was used with six six-plant replications. Experiment 3 was conducted to determine how late in the year plants could be transplanted without suffering significant winter damage.

At approximately 3-day intervals beginning Sept. 6 and ending Oct. 4, 1981, ten 15 cm tall plants growing in peatmoss pots were transplanted directly from the greenhouse to the field.

In all experiments, plant height, number of leaves, and number of basal lateral branches were recorded. Plants in experiments 1 and 2 were measured monthly until Oct. 4, 1981, and again on June 24, 1982, after the first growth flush ceased elongation. Winter survival was assessed in May and June 1982.

All data were subjected to analysis of variance. Mean separation by Duncan's multiple range test was conducted only for effects significant at the 0.05 level.

Results and Discussion

Tissue cultured plants of *Amelanchier laevis* survived transplanting in excellent condition. When results are compared by container type across all other treatments, transplant losses were negligible in either container.

Shoot growth was slightly higher when plants were grown in peat pots, but those transplanted from Cone-Tainers had a higher number of basal lateral branches at the end of the first growing season. Plants exposed to a period of outdoor acclimation under shade prior to field planting appeared to have less transplant shock. Plants acclimated for 5 to 10 days exhibited the greatest incremental increase in height after 1 month.

When the experiment was repeated with an August rather than a June planting date (experiment 2) and using only plants grown in peat pots, a slight benefit to prior acclimation was observed. Increasing time outdoors under shade increased both growth in height and the number of basal lateral shoots observed after 3 months.

The experiment was terminated before the long term effects of acclimation on August transplanting were determined. Survival was 100%. A final study was conducted to determine how late in the season transplanting could be done without affecting overwinter survival (experiment 3). These plants were transplanted to the field directly from the greenhouse as late as Oct. 4, 1982. When survival data were taken on May 12, 1983, it was determined that all transplants had survived the winter and begun active shoot growth.

The number of leaves per plant at transplanting had no effect on transplant or winter survival or on height growth in any of these experiments. It was common for the lower third of the leaves to abscise within 2 weeks of transplanting.

These studies have demonstrated excellent field transplant survival of tissue cultured Amelanchier laevis. Survival levels of nearly 100% can be achieved: 1) without outdoor acclimation under shade, 2) when plants are grown either in peatmoss pots or Cone-Tainers, and 3) at planting dates varying from late June to early October.

A period of outdoor acclimation appears to lessen transplant shock of either June or August transplanting, but the long term (12 month) benefit to this procedure with regard to shoot growth or basal branching appears negligible.

These results should not be generalized to plants which are succulent at the time of transplanting, however. Even though the *Amelanchier* liners used in this study were greenhouse grown, they did have sturdy, woody stems at transplanting.

Micropropagation: Commercial Laboratory Production

Plant Micropropagation using African Violet Leaves

Through the use of biotechnology, desirable genetic traits can be transferred from one organism to another by transfer of DNA. In plants, the DNA of interest is transferred into the new plant by using *Agrobacterium tumefaciens*, a bacterium that can infect plant tissues and incorporate part of its DNA into the DNA of the host plant. Alternatively, a particle gun can be used to directly “shoot” the DNA into plant cells. In both methods, the target for the foreign DNA is a small piece of plant tissue or a small mass of plant cells. Once the DNA has been transferred, new plants must be regenerated from the small pieces of transformed plant tissue using micropropagation (tissue culture) techniques.

Micropropagation is the aseptic culture of cells, pieces of tissue, or organs. It is possible to regenerate new plants from small pieces of plant tissue because each cell of a given plant has the same genetic make-up and is totipotent, that is, capable of developing along a “programmed” pathway leading to the formation of an entire plant that is identical to the plant from which it was derived. In addition to its biotechnological applications, micropropagation is used commercially to asexually propagate plants.

Using micropropagation, millions of new plants can be derived from a single plant. This rapid multiplication allows breeders and growers to introduce new cultivars much earlier than they could by using conventional propagation techniques, such as cuttings. Micropropagation also can be used to establish and maintain virus-free plant stock. This is done by culturing the plant's apical meristem, which typically is not virus-infected, even though the remainder of the plant may be. Once new plants are developed from the apical meristem, they can be maintained and sold as virus-free plants.

Micropropagation differs from all other conventional propagation methods in that aseptic conditions are essential to achieve success. The process of micropropagation can be divided into four stages:

1. Initiation stage. A piece of plant tissue (called an explant) is (a) cut from the plant, (b) disinfested (removal of surface contaminants), and (c) placed on a medium. A medium typically contains mineral salts, sucrose, and a solidifying agent such as agar. The objective of this stage is to achieve an aseptic culture. An aseptic culture is one without contaminating bacteria or fungi.
2. Multiplication stage. A growing explant can be induced to produce vegetative shoots by including a cytokinin in the medium. A cytokinin is a plant growth regulator that promotes shoot formation from growing plant cells.
3. Rooting or preplant stage. Growing shoots can be induced to produce adventitious roots by including an auxin in the medium. Auxins are plant growth regulators that promote root formation. For easily rooted plants, an auxin is usually not necessary and many commercial labs will skip this step.
4. Acclimatization. A growing, rooted shoot can be removed from tissue culture and placed in soil. When this is done, the humidity must be gradually reduced over time because tissue-cultured plants are extremely susceptible to wilting.

Teacher Preparation and Instruction Guide

Preparation for the plant micropropagation exercise should begin at least 24 hours in advance of the laboratory.

The following supplies can be provided to the class in groups of two students:

- 2 fresh African Violet leaves
- 4 culture medium plates
- aseptic work surface/hood
- 1 aluminum foil packet containing 1 sterile forceps
- 1 aluminum foil packet containing 2 sterile razor blades (or 1 sterile #3 scalpel handle with sterile #10 blade attached)

- 2 sterile petri dishes
- 1 litre sterile water
- 1 500-ml or 1-liter jar with screw-cap lid (for disinfesting the leaf)
- alcohol or Bunsen burner
- spray bottle containing 70% ethanol (a spray bottle of the type used to mist plants or spray window cleaning solution works very well).

The following supplies can be shared by four students:

- *500-ml bottle (for sterilizing medium)
- *500-ml or 1-liter plastic or glass beaker (for preparing medium).

The teacher should have available for the entire class:

- commercially-purchased water for medium preparation
- 70% ethanol (dilute 700 ml of 95% ethanol with 250 ml distilled water)
- 0.1% detergent (add 1 ml detergent to 1 litre of water)
- 10% bleach solution plus a few drops of dishwashing detergent
- pH meter or pH paper
- a few mls of the following solutions for adjusting pH:
  - o 1.0 M NaOH
  - o 1.0 M HCl
  - o 0.1 M NaOH
  - o 0.1 M HCl
- parafilm for wrapping plates
- scissors
- Sharpie marking pens
- 5-10 extra sterile scalpel blades
- 5-10 extra sterile petri plates
- dishwashing detergent

Preparation of Plant Tissue Culture Medium

The culture medium is formulated so that it provides all of the compounds needed for plant growth and development, including certain compounds that can be made by an intact plant, but not by an isolated piece of plant tissue. Typically, the medium contains mineral nutrients, organic compounds such as sucrose and vitamins, and plant growth regulators (plant hormones). Agar is added if a solid medium is desired.

Prepare 250 ml per group of four students. In this instruction guide, a prepackaged African violet medium from Carolina Biological Supply Company that includes all necessary components except agar and sucrose, will be used. The instructions that follow are for making 250 ml of medium. Modifications of the procedure needed to make 1 litre are included in brackets ([ ]) in bold.

Step 1. Add about 100 ml [500 ml] of deionized or distilled water to a 500-ml or 1-liter [2-liter] beaker.

Step 2. Add 1.16 grams [one package] of the prepackaged mix to the beaker.

Step 3. Add 7.5 grams [30 grams] of sucrose to the beaker. (Table sugar purchased from a supermarket can be used if reagent-grade sucrose is not available.)

Step 4. Bring the solution up to about 200 ml 00 ml] by adding deionized or distilled water.

Step 5. Stir the solution until all components are dissolved.

Step 6. Adjust the pH to about 5.6 to 5.8 with a pH meter, using NaOH and HCl as necessary. If a pH meter is not available, use pH indicating paper that can detect pH 6.0 ± 0.5 pH units. It is preferable to have the pH at slightly less than 6.0 rather than at slightly greater than 6.0.

At pH < 5.0, the ability of the explant to take up nutrients will be adversely affected. At pH > 7.0, iron will precipitate out of the medium, becoming unavailable for uptake. (When using the prepackaged mix and commercial distilled water, it will take approximately 3 drops [13 drops] of 1.0 NaOH and 6 drops [4 drops] of 0.1 NaOH to raise the pH of 250 ml [1 litre] of medium to 5.6 to 5.8.)

Step 7. Bring the solution up to 250 ml [1 litre] by adding deionized or distilled water, and pour into a 500-ml bottle [1.5 to 2.0-liter bottle] to autoclave.

Step 8. Add 2.0 grams [8.0 grams] agar. (Note: the agar will not dissolve until the medium is heated.)

Step 9. Keep the Cap Loose on the Vessel Containing the Medium while Sterilizing. Sterilize the bottle of medium by placing it in one of the following:

a) boiling water for 30 minutes;
b) a pressure cooker at 15 pounds for 15 minutes; or
c) an autoclave for 15 minutes at 121*f* C (250*f* F) at 1.1 kg/cm^2 (15 psi).

Step 10. After removal from the water bath, pressure cooker, or autoclave, gently swirl the medium to mix the agar. When the agar is completely dissolved and mixed, the medium should appear clear and not turbid.

Step 11. If pouring plates, cool the medium to 45-55*f* C (110*f* F), then pour plates on a clean work surface (spray down with 70% alcohol), working near a small Bunsen burner or alcohol lamp. Use about 25 ml per plate (about one-third to one-half full). If plates are not being used immediately, put them back into the plastic sleeve, tape the sleeve shut, and store plates at 4*f* C (39*f* F).

Sterilization of Supplies

Sterilization of packets of packets of forceps and razor blades can be accomplished by wrapping each item in aluminum foil, labelling the contents with a marking pen, and:

a) baking them in an oven at 350*f* F for 15 minutes;

b) putting them in a pressure cooker at 15 pounds for 15 minutes; or

c) placing them in an autoclave for 15 minutes.

The pressure cooker and autoclave should be at the desired pressure for the 15-minute period. After the packets have cooled, they should be stored unopened at room temperature. The students should be instructed when opening the packets to touch that part that will not come in contact with the plant material or petri dishes. Sterilization of the sterile water needed for the rinses can be accomplished as described for sterilization of the medium.

Note: Teachers may decide for themselves which parts of the procedure can be successfully completed by the student. Middle school students enjoy transferring disinfested explants previously prepared by the instructor to the plates with medium. Older students may be included in parts of the medium preparation and/or the disinfestation procedure. All students will enjoy seeing the growth of shoots from the leaf surface of the African violet. Seeing this regeneration process will make it easier for the teacher to start a discussion of the totipotency of plant cells and the practical application of plant propagation.

Plant Micropropagation

Aseptic Transfer of Cultures to Fresh Medium

Micropropagation is the use of tissue culture methods to propagate plants. Micropropagation if a form of clonal propagation that differs from all other conventional propagation methods in one important aspect: aseptic conditions are necessary to achieve success with micropropagation.

In commercial and research laboratories, aseptic transfer is usually done in a hood that excludes particles larger than 3 microns from the work area. Contaminants of plant tissue culture are bacteria that reside on dust particles and fungi, including spores. A hood will effectively filter out all of these contaminants. It is possible to transfer aseptically without a hood when you take all possible precautions to keep dust and spores from falling in the sterile medium.

One simple, but fairly effective method for accomplishing this, is to spray a smooth counter top with alcohol and spray all other possible sources of contamination (e.g., hands, instruments, and culture vessels) before transferring. A second method used by plant pathologists is to work around a small Bunsen burner, which creates an updraft to prevent spores and dust from falling into the open dish. Extreme caution is required with this method because of the danger of burns.

A third method involves spraying hands, instruments, and vessels with alcohol and working in a large new clear plastic bag. Because newly manufactured plastic is fairly sterile, working inside this sterile bag, which contains still air, will avoid contamination. An old aquarium or cardboard box lined with aluminum foil can be sprayed with alcohol and used to cut down on drafts. With all methods, it is best to keep traffic around the transfer area to a minimum.

Step 1. Put on a pair of sterile gloves. Clean your gloves with 70% ethanol initially, and reclean them if you touch any nonsterile surface or material.

Step 2. Spray your work surface with 70% ethanol.

Step 3. Spray vessels and all tool packets (forceps and razor blades) with 70% ethanol before placing them on the work surface.

Step 4. Arrange the objects on the work surface so that there is a cleared work area for the transferring process. Avoid placing any object "up wind" of your work area.

Step 5. For African violets, use a fresh, healthy leaf. The disinfestation process (removal of surface contaminants) is as follows:

a) Quickly rinse the leaf under cool tap water, then put the whole leaf into a closed container. A jar with a screw-cap lid is best. Depending on the size of the leaves, either a 500- or 1000-ml jar can be used. Wash the leaf in water with 0.1% detergent for

3-4 minutes. Gently agitate the leaf every 20-30 seconds during this washing step.

b) Rinse off the leaf and rinse out the jar with cool tap water.

c) Gently agitate the leaf in 10% bleach solution for 10 minutes. The jar should be filled three-fourths full with the bleach solution. *(The bleach solution disinfests the plant tissue material, killing most fungal and bacterial organisms.)*

d) Pour off the bleach while keeping the lid loosely in place over the container. (Be careful not to spill bleach solution on clothing because it will leave white spots after laundering.) Note: At this point, the leaf is considered sterile. All subsequent rinses should be done with sterile water, and all manipulations of the leaf performed with sterile instruments and supplies. Open one container at a time and never leave the lid off of any container longer than necessary.

e) Take the container to the sterile counter space. Spray the container with 70% ethanol before placing it in the sterile environment. Remove the lid and pour sterile water over the leaves. Fill the jar approximately half-way with the sterile water.

f) Replace the lid and gently agitate for 2-3 minutes to rinse the bleach off the leaves.

g) Pour off the rinse water in the sink, as in Step 5, d.

h) Rinse with sterile water a total of 4 times.

Step 6. Using the sterile razor blade and forceps, cut off bleach-damaged portions in an empty, sterile petri dish. With the empty dish open, hold the leaf with forceps and cut into strips 1.5-cm x 1.5-cm wide. Place 2 or 3 explants into a petri dish that contains the fresh medium. Gently press on the explants with a forceps to ensure that they make contact with the medium. Do two dishes per student.

Step 7. Replace the cover of the petri dish and wrap with parafilm. Use one 1-inch x 4-inch piece per plate. The parafilm will stretch to seal the dish. (Plant tissue cultures require much longer culture times than microbial cultures, so this additional precaution is usually worth the small extra cost and time.)

Step 8. Place the vessel under lights for 16 hours per day if possible. Keep the temperature between 24 and 26*f* C (about 75*f* F). Do not put cultures in direct sunlight, but make sure the area where they are placed is well-lighted by fluorescent bulbs. Growth is slower at lower temperatures.

Step 9. Look for small green shoots forming on or near cut surfaces. It may take 2-3 weeks or more before any visible evidence of new growth is noticeable. You may need a microscope to see these structures as they begin to grow, but after 5-6 weeks, they should be visible to the unaided eye.

If a sterile environment was not maintained, contamination will be obvious within 3 to 4 days. Materials contaminated by fungus will have a fuzzy growth on them, whereas material contaminated by bacteria will have slimy growth on them. Discard contaminated petri dishes promptly to avoid spreading plant diseases to other uncontaminated cultures.

The medium used in this exercise contains a cytokinin, and is specifically formulated to favour the production and multiplication of shoots. Once a sufficient number of shoots has been generated, portions of the explant that contain one or more shoots could be transferred to a medium that contains a higher concentration of an auxin, resulting in root production. Once roots have formed, the plantlets are transferred to pots containing a soil-based or soilless medium, and gradually exposed to conditions of lower humidity and greater light.

Rapid in Vitro Micro propagation of Sugarcane (Saccharum Officinarum L. CV-Nayana) through Callus Culture

Plant tissue culture techniques have become a powerful tool for studying and solving basic and applied problems in plant biotechnology. During the last thirty years, micro propagation and other *in vitro* techniques have become more widely used in commercial horticulture and agriculture for the mass propagation of crop plants. Sugarcane (*Saccharum officinarum* L.) is an important agricultural cash crop in tropical and subtropical region of the world and is the major source of sugar with respect to export product in many developing countries that accounts for more than 60% of the world's sugar production.

It is the only member of the family *Gramineae* belong to genus *Saccharum* in which *in vitro* propagation are standardized and commercially viable. Varieties of sugarcane are highly heterogeneous and generally multiplied vegetatively by stem cutting. In tropical countries nodal sections of sugarcane with 2 or 3 nodes are commonly used as a planting material.

Lack of suitable multiplication procedure and contamination by systemic diseases are the serious problem to multiply an elite genotype of sugarcane in the open field. *In vitro* multiplication of sugarcane has received considerable research attention because of its economic importance as a cash crop. Micro propagation is currently the only realistic means of achieving rapid, large-scale production of disease-free quality planting material as seed canes of newly developed varieties in order to speed up the breeding and commercialization process in Sugarcane. As a result of which plant regeneration through tissue culture technique would be a viable alternative for improving the quality and productivity in sugarcane.

There are reports on tissue culture of sugarcane from different countries but the first attempts to regenerate plants through *in vitro* technique were made on sugarcane by Naz (2003) and Heniz and Mee (1969). Standardization of protocols for *in vitro* multiplication of sugarcane through callus culture, axillary bud and shoot tip culture have been reported by many authors.

How ever, reports are scare on young meristem callus culture in sugarcane cultivar, Nayana of Orissa. The present communication demonstrates an effective high frequency regeneration method which allows for expedient multiplication of micro plants that are easily established ex vitro through callus culture of young meristem as an explant.

Materials and Methods

Explant source: Healthy young meristems were collected by removing the leaf sheath from field grown plants of sugarcane (*Saccharum officinarum* L. cv- Nayana) maintained in the P.G. Department of Botany Utkal University and brought to the laboratory. These young meristems were cut into thin smaller pieces of 1.0 to1.5 cm length. The explants were washed thoroughly under running tap water for 20-30 minutes followed by bavistin 0.2% for 10 minutes and then washed with sterile distilled water and transferred to laminar air flow cabinet. The young meristem explants were treated with 70 % alcohol for 30 second to one minute, followed by another treatment in 0.1% (w/v) mercuric chloride (HgCl2) for another 5 minutes. Finally, the young meristem cuttings were washed thoroughly 3 to 5 times with sterile distilled water before the inoculation in to sterilized nutrient agar media pre-packed in culture tubes. All the above operations were performed under aseptic conditions in laminar airflow cabinet.

Culture medium and condition: The young meristem cutting explants were inoculated on to sterilized semisolid basal MS medium supplemented with different concentrations and combinations of different plant growth regulators.

Callus induction: For callus induction different range of concentration of auxins are tried individually such as 2,4-D (0.5,1.0,2.0,2.5,3.0 and 4.0 mg/l), IBA (0.5,1.0,2.0,2.5,3.0 and 4.0 mg/l) and NAA (0.5,1.0,2.0,2.5,3.0 and 4.0 mg/l) with MS basal medium.

Shoot regeneration Medium: White friable calli were cultured on MS medium supplemented with different combinations and concentrations of BAP, Kinetin (0.5- 4.0mg/l) and IBA, NAA (0.1-1.0mg/l) for multiple shoot regeneration either individually or in combination.

Rooting medium: Elongated micro shoots measuring about 5-6 cm in length were excised from culture tube and transferred to half-strength (1/2 MS) MS medium supplemented with different concentrations of IBA,NAA and IAA (0.5-3.0mg/l) either individually or in combinations.

Environmental condition: The pH of the medium was adjusted to 5.8 before gelling with Agar (6g/l bacteriological grade, Qualigens, India) and prior to autoclaving for 20 min at 120 o C and at 15 lbs psi pressure. Sugar was added at the concentration of 30 gm/l. Molten medium of 20 ml was dispensed into the culture tube and plugged with nonabsorbent cotton wrapped in one layer of cheesecloth. All the cultures were incubated in a growth room with a 16h photoperiod except callus culture (cool, white fluorescent light -30μmol m-2S-1) and the temperature was maintained at 25 ± 30C with 70-80% relative humidity in the culture room. Each treatment consisted of 10 replicates and repeated thrice.

For better callus induction just after autoclaving the culture tube containing semisolid sterilized media should positioned, such a manner as if the inside semisolid media in the culture tube should spread to maximum to wards one side only. At the time of inoculation 2 to 3 explant per culture tube was used. For callus induction experiment culture tubes were kept under 20, 25 and 30°C in complete dark or light/dark condition with 16/8 h light/dark photoperiod with 140μmol m-2S-1 light from cool, white fluorescent lamps.

Acclimatization and transfer of plantlets to soil: Plantlets with well-developed roots were removed from the culture medium. Washed the roots gently under running tap water and were transferred to plastic trays for hardening which contain autoclaved garden soil, farmyard manure and sand (2:1:1). The harden plantlets in the plastic trays were covered with porous polyethylene sheets for maintaining high humidity and were kept under shade in a net house for further growth and development. All were irrigated with 1/8 MS basal salt solution devoid of sucrose and inositol every 4 days for two weeks. After 30 days the plantlets were transplanted in to the soil in field condition.

Statistical Analysis: Experiments were set up in a Randomized Block Design (RBD) and each experiment usually had 10 replicates and was repeated three times. 20 to 30 explants were used per treatment in each replication. Observations were recorded on the percentage of response of callus formation, percentage of response of shoots, number of shoots per callus, shoot length, percentage of response of roots, roots per shoot and root length respectively. The treatment means were compared using Duncan's Multiple Range Test (DMRT) at a 5% probability level according to Gomez and Gomez (1976).

Results and Discussion

Callus Induction: Callus induction was observed within two weeks, after inoculation of the explants on MS medium containing different concentrations of IBA, NAA and 2,4-D (0.5,1.0,2.0,2.5,3.0 and 4.0 mg/l). Although in all concentrations of 2, 4-D the callus induction was triggered, but more profuse callus induction was observed at 2.5 mg/l of 2,4-D with full potential of callus regeneration from the explant of the cultivated varieties, Nayana. On this media composition the explants produced creamy white callus.

The percentage of callus induction was 100 %. Such type of calli has also been reported by Khan et al., (1998) and Khatri et al., (2002). Begum et al., (1995) found that 3.5 mg/l of 2, 4-D produced highest percentage of callus induction from leaf base explant in Bangladesh Nagabari variety of sugarcane. Islam et al., (1982) also reported that 0.5-5.0 mg/l of 2,4-D showed callus induction from leaf tissue on MS medium. The concentrations NAA at 2.0 and 3.0 mg/l produced small amount (20-30%) of callus with grayish globular and hardy in nature. On second sub-culture the calli were turned into non-regenerable callus in NAA combination media with reduced callus weight where as in IBA+ MS basal media composition explant does not showed any remarkable callusing. The cultivated variety Nayana produced poor callus in lower concentration of 2, 4-D and did not at all in 0.5-1.0 mg/l of hormonal combination with MS basal media. All these studies indicated that the sugarcane variety Nayana require higher concentration of 2, 4-D for callus induction.

Regeneration of Micro shoots: Various concentration of cytokinin (BAP and Kinetin) and auxins, (IBA, NAA) were used in different concentration and combinations for shoot regeneration. During this investigation shoot formation was highly influenced by concentrations and type of the growth regulators used in the experiment. Among different concentrations and combinations for shoot multiplication, best performance was showed on MS medium supplemented with BAP (2.0 mg/l) + IBA (0.5mg/l). On this combination 92 % of explant produced shoots.The number of usable shoots was 12.4 ± 1.90 with average length of the shoots 6.2 ± 0.37 cm of the cultivated variety Nayana.

The second best performance was found on MS medium supplemented with BAP (2.0mg/l) + IBA (1.0mg/l) in which average number of usable shoots was 10.5 ± 1.31 with mean length of the shoot 4.0 ± 0.61cm. Islam et al., (1982) also reported the positive effects of BAP+IBA combination on shoot formation in sugarcane. It was also observed that BAP+NAA combination showed effective result but not superior than BAP+IBA combination. However, combinations of high level of cytokinin and a low level of auxin were essential for differentiation of adventitious shoot in sugar cane young meristem callus of the variety Nayana than individual concentration of cytokinin. All these studies concluded that regeneration potential of callus was specific and genotype dependent phenomenon and at the same time it parallel with the hormonal concentration and combinations. It was also observed that callus derived from different auxins showed different regeneration potential. Callus induction, proliferation and regeneration potential in sugarcane exhibited synchrony to each other.

In vitro rooting and acclimatization: Different types of auxins were used at different concentrations and combinations to regenerate adventitious roots. Among them NAA and IBA was found to be comparatively better response than IAA for profuse rooting. NAA+IBA combination showed positive result. Best rooting was observed in 1/2 strength MS medium supplemented with 2.5mg/l NAA and the highest number roots per micro shoots were 13.4 ±1.5, which take only 8-10 days for initiation of root primordial with average root length 4.0 ± 0.94 cm for the variety Nayana found in Orissa. According to Lal and Singh (1994) root can be easily induced on culture shoots by their transfer to another medium with or without NAA,where optimal growth were observed with 1/2 strength of MS medium. Baksha et al., (2002) used 5.0 mg/l NAA for best rooting response in half strength MS medium. Sabaz et al., (2008) used 1.0 mg/l IBA as the best root initiating growth hormone with highest number of 41 roots per plant. Gosal et al.,(1998) obtained rooting on liquid MS medium containing NAA (5 mg/l) and 70 g/l sucrose. Ali and Afghan (2001) observed only 6 - 7 roots after 3 weeks on MS medium containing 2.0 mg/l IBA and 6% sucrose. Baksha et al., (2002) also got rooting response at 0.1 - 0.5 mg/

l IBA along with 0.5 - 2.0 mg/l BAP but these were of poor quality. These findings also agree well with the previous findings of Nadar and Heinz (1977).

Alam et al., (2003) reported best rooting response at 2.5 mg/l IBA with 16 number of roots/explant having 1.1 cm root length. Mamun et al., (2004) obtained best results of rooting on MS medium supplemented with auxins (NAA + IBA) 0.5 mg/l for each one. We also found that 0.5 mg/l NAA+2.5 mg/l IBA was the second best feasible rooting response with 11.3±1.08 number of roots and 3.7 ± 0.47 cm of root length. The plantlets with well developed shoot and roots after acclimatization were successfully transplanted in soil with 85% acclimatization of survivability potential. It is difficult to release a new variety of sugarcane by the conventional breeding method for the genetic behavior of sugar cane.More over also it takes long times to release a stable variety. Thus tissue culture technique can plays an important role in this regard for supply of disease free quality planting material in a year round basis and true to true types of the mother plant.

Chapter 7

Genetic Material: Properties and Replication

DNA

Deoxyribonucleic acid (DNA) is a nucleic acid that contains the genetic instructions used in the development and functioning of all known living organisms with the exception of some viruses. The main role of DNA molecules is the long-term storage of information.

DNA is often compared to a set of blueprints, like a recipe or a code, since it contains the instructions needed to construct other components of cells, such as proteins and RNA molecules. The DNA segments that carry this genetic information are called genes, but other DNA sequences have structural purposes, or are involved in regulating the use of this genetic information.

DNA consists of two long polymers of simple units called nucleotides, with backbones made of sugars and phosphate groups joined by ester bonds. These two strands run in opposite directions to each other and are therefore anti-parallel. Attached to each sugar is one of four types of molecules called bases. It is the sequence of these four bases along the backbone that encodes information. This information is read using the genetic code, which specifies the sequence of the amino acids within proteins.

The code is read by copying stretches of DNA into the related nucleic acid RNA, in a process called transcription. Within cells, DNA is organized into long structures called chromosomes. These chromosomes are duplicated before cells divide, in a process called DNA replication. Eukaryotic organisms (animals, plants, fungi, and protists) store most of their DNA inside the cell nucleus and some of their DNA in organelles, such as mitochondria or chloroplasts. In contrast, prokaryotes (bacteria and archaea) store their DNA only in the cytoplasm. Within the chromosomes, chromatin proteins such as histones compact and organize DNA. These compact structures guide the interactions between DNA and other proteins, helping control which parts of the DNA are transcribed.

Properties

DNA is a long polymer made from repeating units called nucleotides. The DNA chain is 22 to 26 Angströms wide (2.2 to 2.6 nanometres), and one nucleotide unit is 3.3 Å (0.33 nm) long. Although each individual repeating unit is very small, DNA polymers can be very large molecules containing millions of nucleotides. For instance, the largest human chromosome, chromosome number 1, is approximately 220 million base pairs long. In living organisms, DNA does not usually exist as a single molecule, but instead as a pair of molecules that are held tightly together.

These two long strands entwine like vines, in the shape of a double helix. The nucleotide repeats contain both the segment of the backbone of the molecule, which holds the chain together, and a base, which interacts with the other DNA strand in the helix. A base linked to a sugar is called a nucleoside and a base linked to a sugar and one or more phosphate groups is called a nucleotide. If multiple nucleotides are linked together, as in DNA, this polymer is called a polynucleotide.

The backbone of the DNA strand is made from alternating phosphate and sugar residues. The sugar in DNA is 2-deoxyribose, which is a pentose (five-carbon) sugar. The sugars are joined together by phosphate groups that form phosphodiester bonds between the third and fifth carbon atoms of adjacent sugar rings.

These asymmetric bonds mean a strand of DNA has a direction. In a double helix the direction of the nucleotides in one strand is opposite to their direction in the other strand: the strands are *antiparallel*. The asymmetric ends of DNA strands are called the 52 (*five prime*) and 32 (*three prime*) ends, with the 5' end having a terminal phosphate group and the 3' end a terminal hydroxyl group. One major difference between DNA and RNA is the sugar, with the 2-deoxyribose in DNA being replaced by the alternative pentose sugar ribose in RNA.

The DNA double helix is stabilized by hydrogen bonds between the bases attached to the two strands. The four bases found in DNA are adenine (abbreviated A), cytosine (C), guanine (G) and thymine (T). These four bases are attached to the sugar/phosphate to form the complete nucleotide, as shown for adenosine monophosphate.

These bases are classified into two types; adenine and guanine are fused five- and six-membered heterocyclic compounds called purines, while cytosine and thymine are six-membered rings called pyrimidines.

A fifth pyrimidine base, called uracil (U), usually takes the place of thymine in RNA and differs from thymine by lacking a methyl group on its ring. Uracil is not usually found in DNA, occurring only as a breakdown product of cytosine. In addition to RNA and DNA, a large number of artificial nucleic acid analogues have also been created to study the proprieties of nucleic acids, or for use in biotechnology.

Grooves

Twin helical strands form the DNA backbone. Another double helix may be found by tracing the spaces, or grooves, between the strands. These voids are adjacent to the base pairs and may provide a binding site. As the strands are not directly opposite each other, the grooves are unequally sized. One groove, the major groove, is 22 wide and the other, the minor groove, is 12 wide.

The narrowness of the minor groove means that the edges of the bases are more accessible in the major groove. As a result, proteins like transcription factors that can bind to specific sequences in double-stranded DNA usually make contacts to the sides of the bases exposed in the major groove.

This situation varies in unusual conformations of DNA within the cell, but the major and minor grooves are always named to reflect the differences in size that would be seen if the DNA is twisted back into the ordinary B form.

Base Pairing

Each type of base on one strand forms a bond with just one type of base on the other strand. This is called complementary base pairing. Here, purines form hydrogen bonds to pyrimidines, with A bonding only to T, and C bonding only to G. This arrangement of two nucleotides binding together across the double helix is called a base pair. As hydrogen bonds are not covalent, they can be broken and rejoined relatively easily. The two strands of DNA in a double helix can therefore be pulled apart like a zipper, either by a mechanical force or high temperature. As a result of this complementarity, all the information in the double-stranded sequence of a DNA helix is duplicated on each strand, which is vital in DNA replication. Indeed, this reversible and specific interaction between complementary base pairs is critical for all the functions of DNA in living organims.

The two types of base pairs form different numbers of hydrogen bonds, AT forming two hydrogen bonds, and GC forming three hydrogen bonds. DNA with high GC-content is more stable than DNA with low GC-content, but contrary to popular belief, this is not due to the extra hydrogen bond of a GC base pair but rather the contribution of stacking interactions (hydrogen bonding merely provides specificity of the pairing, not stability). As a result, it is both the percentage of GC base pairs and the overall length of a DNA double helix that determine the strength of the association between the two strands of DNA.

Long DNA helices with a high GC content have stronger-interacting strands, while short helices with high AT content have weaker-interacting strands. In biology, parts of the DNA double helix that need to separate easily, such as the TATAAT Pribnow box in some promoters, tend to have a high AT content, making the strands easier to pull apart. In the laboratory, the strength of this interaction can be measured by finding the temperature required to break the hydrogen bonds, their melting temperature (also called T_m value). When all the base pairs in a DNA double helix melt, the strands separate and exist in solution as two entirely independent molecules. These single-stranded DNA molecules (*ssDNA*) have no single common shape, but some conformations are more stable than others.

Sense and Antisense

A DNA sequence is called "sense" if its sequence is the same as that of a messenger RNA copy that is translated into protein. The sequence on the opposite strand is called the "antisense" sequence. Both sense and antisense sequences can exist on different parts of the same strand of DNA (i.e. both strands contain both sense and antisense sequences). In both prokaryotes and eukaryotes, antisense RNA sequences are produced, but the functions of these RNAs are not entirely clear. One proposal is that antisense RNAs are involved in regulating gene expression through RNA-RNA base pairing.

A few DNA sequences in prokaryotes and eukaryotes, and more in plasmids and viruses, blur the distinction between sense and antisense strands by having overlapping genes. In these cases, some DNA sequences do double duty, encoding one protein when read along one strand, and a second protein when read in the opposite direction along the other strand.

In bacteria, this overlap may be involved in the regulation of gene transcription, while in viruses, overlapping genes increase the amount of information that can be encoded within the small viral genome.

Supercoiling

DNA can be twisted like a rope in a process called DNA supercoiling. With DNA in its "relaxed" state, a strand usually circles the axis of the double helix once every 10.4 base pairs, but if the DNA is twisted the strands become more tightly or more loosely wound. If the DNA is twisted in the direction of the helix, this is positive supercoiling, and the bases are held more tightly together. If they are twisted in the opposite direction, this is negative supercoiling, and the bases come apart more easily. In nature, most DNA has slight negative supercoiling that is introduced by enzymes called topoisomerases. These enzymes are also needed to relieve the twisting stresses introduced into DNA strands during processes such as transcription and DNA replication.

Alternate DNA Structures

DNA exists in many possible conformations that include A-DNA, B-DNA, and Z-DNA forms, although, only B-DNA and Z-DNA have been directly observed in functional organisms. The conformation that DNA adopts depends on the hydration level, DNA sequence, the amount and direction of supercoiling, chemical modifications of the bases, the type and concentration of metal ions, as well as the presence of polyamines in solution.

The first published reports of A-DNA X-ray diffraction patterns— and also B-DNA used analyses based on Patterson transforms that provided only a limited amount of structural information for oriented fibres of DNA. An alternate analysis was then proposed by Wilkins *et al.*, in 1953, for the *in vivo* B-DNA X-ray diffraction/scattering patterns of highly hydrated DNA fibres in terms of squares of Bessel functions. In the same journal, James D. Watson and Francis Crick presented their molecular modeling analysis of the DNA X-ray diffraction patterns to suggest that the structure was a double-helix.

Although the 'B-DNA form' is most common under the conditions found in cells, it is not a well-defined conformation but a family of related DNA conformations that occur at the high hydration levels present in living cells. Their corresponding X-ray diffraction and scattering patterns are characteristic of molecular paracrystals with a significant degree of disorder.

Compared to B-DNA, the A-DNA form is a wider right-handed spiral, with a shallow, wide minor groove and a narrower, deeper major groove. The A form occurs under non-physiological conditions in partially dehydrated samples of DNA, while in the cell it may be produced in hybrid pairings of DNA and RNA strands, as well as in enzyme-DNA complexes. Segments of DNA where the bases have been chemically modified by methylation may undergo a larger change in conformation and adopt the Z form. Here, the strands turn about the helical axis in a left-handed spiral, the opposite of the more common B form. These unusual structures can be recognized by specific Z-DNA binding proteins and may be involved in the regulation of transcription.

Quadruplex Structures

At the ends of the linear chromosomes are specialized regions of DNA called telomeres. The main function of these regions is to allow the cell to replicate chromosome ends using the enzyme telomerase, as the enzymes that normally replicate DNA cannot copy the extreme 32 ends of chromosomes. These specialized chromosome caps also help protect the DNA ends, and stop the DNA repair systems in the cell from treating them as damage to be corrected. In human cells, telomeres are usually lengths of single-stranded DNA containing several thousand repeats of a simple TTAGGG sequence.

These guanine-rich sequences may stabilize chromosome ends by forming structures of stacked sets of four-base units, rather than the usual base pairs found in other DNA molecules. Here, four guanine bases form a flat plate and these flat four-base units then stack on top of each other, to form a stable *G-quadruplex* structure.

These structures are stabilized by hydrogen bonding between the edges of the bases and chelation of a metal ion in the centre of each four-base unit. Other structures can also be formed, with the central set of four bases coming from either a single strand folded around the bases, or several different parallel strands, each contributing one base to the central structure.

In addition to these stacked structures, telomeres also form large loop structures called telomere loops, or T-loops. Here, the single-stranded DNA curls around in a long circle stabilized by telomere-binding proteins. At the very end of the T-loop, the single-stranded telomere DNA is held onto a region of double-stranded DNA by the telomere strand disrupting the double-helical DNA and base pairing to one of the two strands. This triple-stranded structure is called a displacement loop or D-loop.

Branched DNA

In DNA fraying occurs when non-complementary regions exist at the end of an otherwise complementary double-strand of DNA. However, branched DNA can occur if a third strand of DNA is introduced and contains adjoining regions able to hybridize with the frayed regions of the pre-existing double-strand. Although the simplest example of branched DNA involves only three strands of DNA, complexes involving additional strands and multiple branches are also possible.

Vibration

DNA may carry out low-frequency collective motion as observed by the Raman spectroscopy and analyzed with a quasi-continuum model.

Chemical Modifications

Base Modifications

The expression of genes is influenced by how the DNA is packaged in chromosomes, in a structure called chromatin. Base modifications can be involved in packaging, with regions that have low or no gene expression usually containing high levels of methylation of cytosine bases. For example, cytosine methylation, produces 5-methylcytosine, which is important for X-chromosome inactivation.

The average level of methylation varies between organisms - the worm *Caenorhabditis elegans* lacks cytosine methylation, while vertebrates have higher levels, with up to 1% of their DNA containing 5-methylcytosine. Despite the importance of 5-methylcytosine, it can deaminate to leave a thymine base, so methylated cytosines are particularly prone to mutations. Other base modifications include adenine methylation in bacteria, the presence of 5-hydroxymethylcytosine in the brain, and the glycosylation of uracil to produce the "J-base" in kinetoplastids.

Damage

DNA can be damaged by many sorts of mutagens, which change the DNA sequence. Mutagens include oxidizing agents, alkylating agents and also high-energy electromagnetic radiation such as ultraviolet light and X-rays. The type of DNA damage produced depends on the type of mutagen. For example, UV light can damage DNA by producing thymine dimers, which are cross-links between pyrimidine bases.

On the other hand, oxidants such as free radicals or hydrogen peroxide produce multiple forms of damage, including base modifications, particularly of guanosine, and double-strand breaks. A typical human cell contains about 150,000 bases that have suffered oxidative damage. Of these oxidative lesions, the most dangerous are double-strand breaks, as these are difficult to repair and can produce point mutations, insertions and deletions from the DNA sequence, as well as chromosomal translocations.

Many mutagens fit into the space between two adjacent base pairs, this is called *intercalation*. Most intercalators are aromatic and planar molecules; examples include ethidium bromide, daunomycin, and doxorubicin. In order for an intercalator to fit between base pairs, the bases must separate, distorting the DNA strands by unwinding of the double helix. This inhibits both transcription and DNA replication, causing toxicity and mutations.

As a result, DNA intercalators are often carcinogens, and benzo[*a*]pyrene diol epoxide, acridines, aflatoxin and ethidium bromide are well-known examples. Nevertheless, due to their ability to inhibit DNA transcription and replication, other similar toxins are also used in chemotherapy to inhibit rapidly growing cancer cells.

Biological Functions

DNA usually occurs as linear chromosomes in eukaryotes, and circular chromosomes in prokaryotes. The set of chromosomes in a cell makes up its genome; the human genome has approximately 3 billion base pairs of DNA arranged into 46 chromosomes. The information carried by DNA is held in the sequence of pieces of DNA called genes. Transmission of genetic information in genes is achieved via complementary base pairing.

For example, in transcription, when a cell uses the information in a gene, the DNA sequence is copied into a complementary RNA sequence through the attraction between the DNA ad the correct RNA nucleotides. Usually, this RNA copy is then used to make a matching protein sequence in a process called translation, which depends on the same interaction between RNA nucleotides.

In alternative fashion, a cell may simply copy its genetic information in a process called DNA replication. The details of these functions are covered in other articles; here we focus on the interactions between DNA and other molecules that mediate the function of the genome.

Genes and Genomes

Genomic DNA is tightly and orderly packed in the process called DNA condensation to fit the small available volumesof the cell. In eukaryotes, DNA is located in the cell nucleus, as well as small amounts in mitochondria and chloroplasts. In prokaryotes, the DNA is held within an irregularly shaped body in the cytoplasm called the nucleoid. The genetic information ina genome is held within genes, and the complete set of this information in an organism is called its genotype. A gene is a unit of heredity and is a region of DNA that influences a particular characteristic in an organism. Genes contain an open reading frame that can be transcribed, as well as regulatory sequences such as promoters and enhancers, which control the transcription of the open reading frame.

In many species, only a small fraction of the total sequence of the genome encodes protein. For example, only about 1.5% of the human genome consists of protein-coding exons, with over 50% of human DNA consisting of non-coding repetitive sequences. The reasons for the presence of so much non-coding DNA in eukaryotic genomes and the extraordinary differences in genome size, or *C-value*, among species represent a long-standing puzzle known as the "C-value enigma". However, DNA sequences that do not code protein may still encode functional non-coding RNA molecules, which are involved in the regulation of gene expression.

Some non-coding DNA sequences play structural roles in chromosomes. Telomeres and centromeres typically contain few genes, but are important for the function and stability of chromosomes. An abundant form of non-coding DNA in humans are pseudogenes, which are copies of genes that have been disabled by mutation. These sequences are usually just molecular fossils, although they can occasionally serve as raw genetic material for the creation of new genes through the process of gene duplication and divergence.

Transcription and Translation

A gene is a sequence of DNA that contains genetic information and can influence the phenotype of an organism. Within a gene, the sequence of bases along a DNA strand defines a messenger RNA sequence, which then defines one or more protein sequences.

The relationship between the nucleotide sequences of genes and the amino-acid sequences of proteins is determined by the rules of translation, known collectively as the genetic code. The genetic code consists of three-letter 'words' called *codons* formed from a sequence of three nucleotides (e.g. ACT, CAG, TTT). In transcription, the codons of a gene are copied into messenger RNA by RNA polymerase. This RNA copy is then decoded by a ribosome that reads the RNA sequence by base-pairing the messenger RNA to transfer RNA, which carries amino acids. Since there are 4 bases in 3-letter combinations, there are 64 possible codons (4 combinations). These encode the twenty standard amino acids, giving most amino acids more than one possible codon. There are also three 'stop' or 'nonsense' codons signifying the end of the coding region; these are the TAA, TGA and TAG codons.

Replication

Cell division is essential for an organism to grow, but, when a cell divides, it must replicate the DNA in its genome so that the two daughter cells have the same genetic information as their parent. The double-stranded structure of DNA provides a simple mechanism for DNA replication. Here, the two strands are separated and then each strand's complementary DNA sequence is recreated by an enzyme called DNA polymerase. This enzyme makes the complementary strand by finding the correct base through complementary base pairing, and bonding it onto the original strand. As DNA polymerases can only extend a DNA strand in a 52 to 32 direction, different mechanisms are used to copy the antiparallel strands of the double helix. In this way, the base on the old strand dictates which base appears on the new strand, and the cell ends up with a perfect copy of its DNA.

Interactions with Proteins

All the functions of DNA depend on interactions with proteins. These protein interactions can be non-specific, or the protein can bind specifically to a single DNA sequence. Enzymes can also bind to DNA and of these, the polymerases that copy the DNA base sequence in transcription and DNA replication are particularly important.

DNA-binding Proteins

Structural proteins that bind DNA are well-understood examples of non-specific DNA-protein interactions. Within chromosomes, DNA is held in complexes with structural proteins. These proteins organize the DNA into a compact structure called chromatin. In eukaryotes this structure involves DNA binding to a complex of small basic proteins called histones, while in prokaryotes multiple types of proteins are involved. The histones form a disk-shaped complex called a nucleosome, which contains two complete turns of double-stranded DNA wrapped around its surface.

These non-specific interactions are formed through basic residues in the histones making ionic bonds to the acidic sugar-phosphate backbone of the DNA, and are therefore largely independent of the base sequence. Chemical modifications of these basic amino acid residues include methylation, phosphorylation and acetylation. These chemical changes alter the strength of the interaction between the DNA and the histones, making the DNA more or less accessible to transcription factors and changing the rate of transcription. Other non-specific DNA-binding proteins in chromatin include the high-mobility group proteins, which bind to bent or distorted DNA. These proteins are important in bending arrays of nucleosomes and arranging them into the larger structures that make up chromosomes.

A distinct group of DNA-binding proteins are the DNA-binding proteins that specifically bind single-stranded DNA. In humans, replication protein A is the best-understood member of this family and is used in processes where the double helix is separated, including DNA replication, recombination and DNA repair. These binding proteins seem to stabilize single-stranded DNA and protect it from forming stem-loops or being degraded by nucleases. In contrast, other proteins have evolved to bind to particular DNA sequences. The most intensively studied of these are the various transcription factors, which are proteins that regulate transcription. Each transcription factor binds to one particular set of DNA sequences and activates or inhibits the transcription of genes that have these sequences close to their promoters. The transcription factors do this in two ways. Firstly, they can bind the RNA polymerase responsible for transcrption, either directly or through other mediator proteins; this locates the polymerase at the promoter and allows it to begin transcription. Alternatively, transcription factors can bind enzymes that modify the histones at the promoter; this will change the accessibility of the DNA template to the polymerase.

As these DNA targets can occur throughout an organism's genome, changes in the activity of one type of transcription factor can affect thousands of genes. Consequently, these proteins are often the targets of the signal transduction processes that control responses to environmental changes or cellular differentiation and development. The specificity of these transcription factors' interactions with DNA come from the proteins making multiple contacts to the edges of the DNA bases, allowing them to "read" the DNA sequence. Most of these base-interactions are made in the major groove, where the bases are most accessible.

DNA-modifying Enzymes

Nucleases and Ligases

Nucleases are enzymes that cut DNA strands by catalyzing the hydrolysis of the phosphodiester bonds. Nucleases that hydrolyse nucleotides from the ends of DNA strands are called exonucleases, while endonucleases cut within strands. The most frequently used nucleases in molecular biology are the restriction endonucleases, which cut DNA at specific sequences. For instance, the EcoRV enzyme shown to the left recognizes the 6-base sequence 52 -GAT | ATC-32 and makes a cut at the vertical line. In nature, these enzymes protect bacteria against phage infection by digesting the phage DNA when it enters the bacterial cell, acting as art of the restriction modification system. In technology, these sequence-speific nucleases are used in molecular cloning and DNA fingerprinting.

Enzymes called DNA ligases can rejoin cut or broken DNA strands. Ligases are particularly important in lagging strand DNA replication, as they join together the short segments of DNA produced at the replication fork into a complete copy of the DNA template. They are also used in DNA repair and genetic recombination.

Topoisomerases and Helicases

Topoisomerases are enzymes with both nuclease and ligase activity. These proteins change the amount of supercoiling in DNA. Some of these enzymes work by cutting the DNA helix and allowing one section to rotate, thereby reducing its level of supercoiling; the enzyme then seals the DNA break. Other types of these enzymes are capable of cutting one DNA helix and then passing a second strand of DNA through this break, before rejoining the helix. Topoisomerases are required for many processes involving DNA, such as DNA replication and transcription. Helicses are proteins that are a type of molecular motor. They use the chemical energy in nucleoside triphosphates, predominantly ATP, to break hydrogen bonds between bases and unwind the DNA double helix into single strands. These enzymes are essential for most processes where enzymes need to access the DNA bases.

Polymerases

Polymerases are enzymes that synthesize polynucleotide chains from nucleoside triphosphates. The sequence of their products are copies of existing polynucleotide chains - which are called *templates*. These enzymes function by adding nucleotides onto the 32 hydroxyl group of the previous nucleotide in a DNA strand. As a consequence, all polymerases work in a 52 to 32 direction. In the active site of these enzymes, the incoming nucleoside triphosphate base-pairs to the template: this allows polymerases to accurately synthesize the complementary strand of their template. Polymerases are classified according to the type of template that they use.

In DNA replication, a DNA-dependent DNA polymerase makes a copy of a DNA sequence. Accuracy is vital in this process, so many of these polymerases have a proofreading activity. Here, the polymerase recognizes the occasional mistakes in the synthesis reaction by the lack of base pairing between the mismatched nucleotides. If a mismatch is detected, a 32 to 52 exonuclease activity is activated and the incorrect base removed. In most organisms, DNA polymerases function in a large complex called the replisome that contains multiple accessory subunits, such as the DNA clamp or helicases.

RNA-dependent DNA polymerases are a specialized class of polymerases that copy the sequence of an RNA strand into DNA. They include reverse transcriptase, which is a viral enzyme involved in the infection of cells by retroviruses, and telomerase, which is required for the replication of telomeres. Telomerase is an unusual polymerase because it contains its own RNA template as part of its structure.

Transcription is carried out by a DNA-dependent RNA polymerase that copies the sequence of a DNA strand into RNA. To begin transcribing a gene, the RNA polymerase binds to a sequence of DNA called a promoter and separates the DNA strands. It then copies the gene sequence into a messenger RNA transcript until it reaches a region of DNA called the terminator, where it halts and detaches from the DNA. As with human DNA-dependent DNA polymerases, RNA polymerase II, the enzyme that transcribes most of the genes in the human genome, operates as part of a large protein complex with multiple regulatory and accessory subunits.

Genetic Recombination

A DNA helix usually does not interact with other segments of DNA, and in human cells the different chromosomes even occupy separate areas in the nucleus called "chromosome territories". This physical separation of different chromosomes is important for the ability of DNA to function as a stable repository for information, as one of the few times chromosomes interact is during chromosomal crossover when they recombine. Chromosomal crossover is when two DNA helices break, swap a setion and then rejoin.

Recombination allows chromosomes to exchange genetic information and produces new combinations of genes, which increases the efficiency of natural selection and can be important in the rapid evolution of new proteins. Genetic recombination can also be involved in DNA repair, particularly in the cell's response to double-strand breaks. The most common form of chromosomal crossover is homologous recombination, where the two chromosomes involved share very similar sequences. Non-homologous recombination can be damaging to cells, as it can produce chromosomal translocations and genetic abnormalities. The recombination reaction is catalyzed by enzymes known as recombinses, such as RA51.

The first step in recombination is a double-stranded break either caused by an endnuclease or damage to the DNA. A series of steps catalyzed in part by the recombinase then leads to joining of the two helices by at least one Holliday junction, in which a segment of a single strand in each helix is annealed to the complementary strand in the other helix. The Holliday junction is a tetrahedral junction structure that can be moved along the pair of chromosomes, swapping one strand for another. The recombination reaction is then halted by cleavage of the junction and re-ligation of the released DNA.

Evolution

DNA contains the genetic information that allows all modern living things to function, grow and reproduce. However, it is unclear how long in the 4-billion-year history of life DNA has performed this function, as it has been proposed that the earliest forms of life may have used RNA as their genetic material. RNA may have acted as the central part of early cell metabolism as it can both transmit genetic information and carry out catalysis as part of ribozymes. This ancient RNA world where nucleic acid would have been used for both catalysis and genetics may have influenced the evolution of the current genetic code based on four nucleotide bases. This would occur, since the number of different bases in such an organism is a trade-off between a small number of bases increasing replication accuracy and a large number of bases increasing the catalytic efficiency of ribozymes. However, there is no direct evidence of ancient genetic systems, as recovery of DNA from most fossils is impossible. This is because DNA will survive in the environment for less than one million years and slowly degrades into short fragments in solution. Claims for older DNA have been made, most noably a report of the isolation of a viable bacterium from a salt crystal 250 million years old, but these claims are controversial.

Uses in Technoloy

Genetic Engineering

Methods have been developed to purify DNA from organisms, such as phenol-chloroform extraction and manipulate it in the laboratory, such as restriction digests and the polymerase chain reaction. Modern biology and biochemistry make intensive use of these techniques in recombinan DNA technolog. Recombinant DNA is a man-made DNA sequence that has been assembled from other DNA sequences. They can be transormed into organisms in the form of plasmids or in the appropriate format, by using a viral vector. The genetically modified organisms produced can be used to produce products such as recombinant proteins, used in medical research, or be grown in agriculture.

Forensics

Forensic scientists can use DNA in blood, semen, skin, saliva or hair found at a crime scene to identify a matching DNA of an individual, such a a perpetrator. This process is formally termed DNA profiling, but may also be called "genetic fingerprinting". In DNA profiling, the lengths of variable sections of repetitive DNA, such as short tandem repeats and minisatellites, are compared between people.

This methodis usually an extremely reliable technique for identifying a matching DNA. However, identification can be complicated if the scene is contaminated with DNA from several people. DNA profling was developed in 1984 by British geneticist Sir Alec Jeffreys, and first used in forensic science to convict Colin Pitchfork in the 1988 Enderby murders case.

People convicted of certain types of crimes may be required to provide a sample of DNA for a database. This has helped investigators solve old cases where only a DNA sample was obtained from the scene. DNA profiling can also be used to identify victims of mass casualty incidents. On the other hand, many convicted people have been released from prison on the basis of DNA techniques, which were not available when a crime had originally been committed.

DNA Profiling

DNA profiling (also called DNA testing, DNA typing, or genetic fingerprinting) is a technique employed by forensic scientists to assist in the identification of individuals on the basis of their respective DNA profiles. DNA profiles are encrypted sets of numbers that reflect a person's DNA make-up, which can also be used as the person's identifier.

DNA profiling should not be confused with full genome sequencing. It is used in, for example, parental testing and rape investigation.

Although 99.9% of human DNA sequences are the same in every person, enough of the DNA is different to distinguish one individual from another. DNA profiling uses repetitive ("repeat") sequences that are highly variable, called variable number tandem repeats (VNTR). VNTRs loci are very similar between closely related humans, but so variable that unrelated individuals are extremely unlikely to have the same VNTRs. The DNA profiling technique was first reported in 1984 by Sir Alec Jeffreys at the University of Leicester in England, and is now the basis of several national DNA databases. Dr. Jeffreys's genetic fingerprinting was made commercially available in 1987, when a chemical company, ICI, started a blood-testing centre in England.

DNA Profiling Process

The process begins with a sample of an individual's DNA (typically called a "reference sample"). The most desirable method of collecting a reference sample is the use of a buccal swab, as this reduces the possibility of contamination. When this is not available (e.g. because a court order may be needed and not obtainable) other methods may need to be used to collect a sample of blood, saliva, semen, or other appropriate fluid or tissue from personal items (e.g. toothbrush, razor, etc.) or from stored samples (e.g. banked sperm or biopsy tissue). Samples obtained from blood relatives (biological relative) can provide an indication of an individual's profile, as could human remains which had been previously profiled. A reference sample is then analyzed to create the individual's DNA profile using one of a number of techniques, discussed below. The DNA profile is then compared against another sample to determine whether there is a genetic match.

RFLP Analysis

The first methods for finding out genetics used for DNA profiling involved restriction enzyme digestion, followed by Southern blot analysis. Although polymorphisms can exist in the restriction enzyme cleavage sites, more commonly the enzymes and DNA probes were used to analyze VNTR loci. However, the Southern blot technique is laborious, and require large amounts of undegraded sample DNA. Also, Karl Brown's original technique looked at many minisatellite loci at the same time, increasing the observed variability, but making it hard to discern individual alleles (and thereby precluding parental testing). These early techniques have been supplanted by PCR-based assays.

PCR Analysis

With the invention of the polymerase chain reaction (PCR) technique, DNA profiling took huge strides forward in both discriminating power and the ability to recover information from very small (or degraded) starting samples. PCR greatly amplifies the amounts of a specific region of DNA, using oligonucleotide primers and a thermostable DNA polymerase.

Early assays such as the HLA-DQ alpha reverse dot blot strips grew to be very popular due to their ease of use, and the speed with which a result could be obtained. However they were not as discriminating as RFLP. It was also difficult to determine a DNA profile for mixed samples, such as a vaginal swab from a sexual assault victim.

Fortunately, the PCR method is readily adaptable for analyzing VNTR loci. In the United States the FBI has standardized a set of 13 VNTR assays for DNA typing, and has organized the CODIS database for forensic identification in criminal cases. Similar assays and databases have been set up in other countries. Also, commercial kits are available that analyze single-nucleotide polymorphisms (SNPs). These kits use PCR to amplify specific regions with known variations and hybridize them to probes anchored on cards, which results in a coloured spot corresponding to the particular sequence variation.

STR Analysis

The method of DNA profiling used today is based on PCR and uses short tandem repeats (STR). This method uses highly polymorphic regions that have short repeated sequences of DNA (the most common is 4 bases repeated, but there are other lengths in use, including 3 and 5 bases).

Because unrelated people almost certainly have different numbers of repeat units, STRs can be used to discriminate between unrelated individuals. These STR loci (locations on a chromosome) are targeted with sequence-specific primers and amplified using PCR. The DNA fragments that result are then separated and detected using electrophoresis. There are two common methods of separation and detection, capillary electrophoresis (CE) and gel electrophoresis.

Each STR is polymorphic, however, the number of alleles is small. Typically each STR allele will be shared by around 5 - 20% of individuals. The power of STR analysis comes from looking at multiple STR loci simultaneously. The pattern of alleles can identify an individual quite accurately. Thus STR analysis provides an excellent identification tool.

The more STR regions that are tested in an individual the more discriminating the test becomes.

From country to country, different STR-based DNA-profiling systems are in use. In North America, systems which amplify the CODIS 13 core loci are almost universal, while in the UK the SGM+ system (which is compatible with The National DNA Database), is in use. Whichever system is used, many of the STR regions used are the same. These DNA-profiling systems are based on multiplex reactions, whereby many STR regions will be tested at the same time.

The true power of STR analysis is in its statistical power of discrimination. Because the 13 loci that are currently used for discrimination in CODIS are independently assorted (having a certain number of repeats at one locus doesn't change the likelihood of having any number of repeats at any other locus), the product rule for probabilities can be applied.

This means that if someone has the DNA type of ABC, where the three loci were independent, we can say that the probability of having that DNA type is the probability of having type A times the probability of having type B times the probability of having type C. This has resulted in the ability to generate match probabilities of 1 in a quintillion (1 with 18 zeros after it) or more. However, DNA database searches showed much more frequent than expected false DNA matches including one perfect 13 locus match out of only 30,000 DNA samples in Maryland in January 2007.

Moreover, since there are about 12 million monozygotic twins on Earth, that theoretical probability is useless. For example, the actual probability that 2 random people have the same DNA depends on whether there were twins or triplets (etc.) in the family, and the number of loci used in the test. Where twins are common, the probability of matching the DNA is 22 in 1000, or about 2.2 in 100 will have matching DNA.

In practice, the risk of contaminated-matching is much greater than matching a distant relative, such as a sample being contaminated from nearby objects, or from left-over cells transferred from a prior test. Logically, the risk is greater for matching the most common person in the samples: everything collected from, or in contact with, a victim is a major source of contamination for any other samples brought into a lab. For that reason, multiple control-samples are typically tested, to ensure that they stayed clean, when prepared during the same period as the actual test samples. Unexpected matches (or variations) in several control-samples indicates a high probability of contamination for the actual test samples. In a relationship test, the full DNA profiles should differ (except for twins), to prove that a person wasn't actually matched as being related to their own DNA in another sample.

AmpFLP

Another technique, AmpFLP, or amplified fragment length polymorphism was also put into practice during the early 1990s. This technique was also faster than RFLP analysis and used PCR to amplify DNA samples. It relied on variable number tandem repeat (VNTR) polymorphisms to distinguish various alleles, which were separated on a polyacrylamide gel using an allelic ladder (as opposed to a molecular weight ladder). Bands could be visualized by silver staining the gel. One popular locus for fingerprinting was the D1S80 locus.

As with all PCR based methods, highly degraded DNA or very small amounts of DNA may cause allelic dropout (causing a mistake in thinking a heterozygote is a homozygote) or other stochastic effects. In addition, because the analysis is done on a gel, very high number repeats may bunch together at the top of the gel, making it difficult to resolve. AmpFLP analysis can be highly automated, and allows for easy creation of phylogenetic trees based on comparing individual samples of DNA. Due to its relatively low cost and ease of set-up and operation, AmpFLP remains popular in lower income countries.

DNA Family Relationship Analysis

Using PCR technology, DNA analysis is widely applied to determine genetic family relationships such as paternity, maternity, siblingship and other kinships.

During conception, the father's sperm cell and the mother's egg cell, each containing half the amount of DNA found in other body cells, meet and fuse to form a fertilized egg, called a zygote. The zygote contains a complete set of DNA molecules, a unique combination of DNA from both parents. This zygote divides and multiplies into an embryo and later, a full human being. At each stage of development, all the cells forming the body contain the same DNA—half from the father and half from the mother. This fact allows the relationship testing to use all types of all samples including loose cells from the cheeks collected using buccal swabs, blood or other types of samples.

While a lot of DNA contains information for a certain function, there is some called junk DNA, which is currently used for human identification. At some special locations (called loci) in the junk DNA, predictable inheritance patterns were found to be useful in determining biological relationships. These locations contain specific DNA markers that DNA scientists use to identify individuals.

In a routine DNA paternity test, the markers used are Short Tandem Repeats (STRs), short pieces of DNA that occur in highly differential repeat patterns among individuals. Each person's DNA contains two copiesof these markers—one copy inherited from the father and one from the mother.

Within a population, the markers at each person's DNA location could differ in length and sometimes sequence, depending on the markers inherited from the parents. The combination of marker sizes found in each person makes up his/her unique genetic profile. When determining the relationship between two individuals, their genetic profiles are compared to see if they share the same inheritance patterns at a statistically conclusive rate.

For example, the following sample report from this commercial DNA paternity testing laboratory Universal Genetics signifies how relatedness between parents and child is identified on those special markers:

DNA Marker	***Mother***	***Child***	***Alleged father***
D21S11	28, 30	28, 31	29, 31
D7S820	9, 10	10, 11	11, 12
TH01	14, 15	14, 16	15, 16
D13S317	7, 8	7, 9	8, 9
D19S433	14, 16.2	14, 15	15, 17

The partial results indicate that the child and the alleged father's DNA match among these five markers. The complete test results show this correlation on 16 markers between the child and the tested man to draw a conclusion of whether or not the man is the biological father.

Scientifically, each marker is assigned with a Paternity Index (PI), which is a statistical measure of how powerfully a match at a particular marker indicates paternity.

The PI of each marker is multiplied with each other to generate the Combined Paternity Index (CPI), which indicates the overall probability of an individual being the biological father of the tested child relative to any random man from the entire population of the same race. The CPI is then converted into a Probability of Paternity showing the degree of relatedness between the alleged father and child.

The DNA test report in other family relationship tests, such as grandparentage and siblingship tests, is similar to a paternity test report. Instead of the Combined Paternity Index, a different value, such as a Siblingship Index, is reported. The report shows the genetic profiles of each tested person. If there are markers shared among the tested individuals, the probability of biological relationship is calculated to determine how likely the tested individuals share the same markers due to a blood relationship.

Y-chromosome Analysis

Recent innovations have included the creation of primers targeting polymorphic regions on the Y-chromosome (Y-STR), which allows resolution of a mixed DNA sample from a male and female and/or cases in which a differential extraction is not possible. Y-chromosomes are paternally inherited, so Y-STR analysis can help in the identification of paternally related males. Y-STR analysis was performed in the Sally Hemings controversy to determine if Thomas Jefferson had sired a son with one of his slaves.

Mitochondrial Analysis

For highly degraded samples, it is sometimes impossible to get a complete profile of the 13 CODIS STRs. In these situations, mitochondrial DNA (mtDNA) is sometimes typed due to there being many copies of mtDNA in a cell, while there may only be 1-2 copies of the nuclear DNA. Forensic scientists amplify the HV1 and HV2 regions of the mtDNA, then sequence each region and compare single-nucleotide differences to a reference. Because mtDNA is maternally inherited, directly linked maternal relatives can be used as match references, such as one's maternal grandmother's daughter's son.

A difference of two or more nucleotides is generally considered to be an exclusion. Heteroplasmy and poly-C differences may throw off straight sequence comparisons, so some expertise on the part of the analyst is required. mtDNA is useful in determining clear identities, such as those of missing people when a maternally linked relative can be found. mtDNA testing was used in determining that Anna Anderson was not the Russian princess she had claimed to be, Anastasia Romanov.

mtDNA can be obtained from such material as hair shafts and old bones/teeth..

DNA Databases

There are now several DNA databases in existence around the world. Some are private, but most of the largest databases are government controlled. The United States maintains the largest DNA database, with the Combined DNA Index System, holding over 5 million records as of 2007.

The United Kingdom maintains the National DNA Database (NDNAD), which is of similar size, despite the UK's smaller population. The size of this database, and its rate of growth, is giving concern to civil liberties groups in the UK, where police have wide-ranging powers to take samples and retain them even in the event of acquittal.

The U.S. Patriot Act of the United States provides a means for the U.S. government to get DNA samples from other countries if they are either a division of, or head office of, a company operating in the U.S.

Under the act, the American offices of the company can't divulge to their subsidiaries/offices in other countries the reasons that these DNA samples are sought or by whom.

When a match is made from a National DNA Databank to link a crime scene to an offender who has provided a DNA Sample to a databank that link is often referred to as a *cold hit*. A cold hit is of value in referring the police agency to a specific suspect but is of less evidential value than a DNA match made from outside the DNA Databank.

Considerations when Evaluating DNA Evidence

In the early days of the use of genetic fingerprinting as criminal evidence, juries were often swayed by spurious statistical arguments by defence lawyers along these lines: given a match that had a 1 in 5 million probability of occurring by chance, the lawyer would argue that this meant that in a country of say 60 million people there were 12 people who woul also match the profile. This was then translated to a 1 in 12 chance of the suspect being the guilty one.

This argument is not sound unless the suspect was drawn at random from the population of the country. In fact, a jury should consider how likely it is that an individual matching the genetic profile would also have been a suspect in the case for other reasons. Another spurious statistical argument is based on the false assumption that a 1 in 5 million probability of a match automatically translates into a 1 in 5 million probability of innocence and is known as the prosecutor's fallacy.

When using RFLP, the theoretical risk of a coincidental match is 1 in 100 billion (100,000,000,000), although the practical risk is actually 1 in 1000 because monozygotic twins are 0.2% of the human population. Moreover, the rate of laboratory error is almost certainly higher than this, and often actual laboratory procedures do not reflect the theory under which the coincidence probabilities were computed.

For example, the coincidence probabilities may be calculated based on the probabilities that markers in two samples have bands in *precisely* the same location, but a laboratory worker may conclude that similar—but not precisely identical—band patterns result from identical genetic samples with some imperfection in the agarose gel. However, in this case, the laboratory worker increases the coincidence risk by expanding the criteria for declaring a match. Recent studies have quoted relatively high error rates which may be cause for concern.

In the early days of genetic fingerprinting, the necessary population data to accurately compute a match probability was sometimes unavaiable. Between 1992 and 1996, arbitrary low ceilings were controversially put on match probabilities used in RFLP analysis rather than the higher theoretically computedones.

Today, RFLP has become widely disused due to the advent of more discriminating, sensitive and easier technologies. STRs do not suffer from such subjectivity and provide similar power of discrimination (1 in 10^13 for unrelated individuals if using a full SGM+ profile) It should be noted that figures of this magnitude are not considered to be statistically supportable by scientists in the UK, for unrelated individuals with full matching DNA profiles a match probability of 1 in a billion is considered statistically supportable (Since 1998 the DNA profiling system supported by The National DNA Database in the UK is the SGM+ DNA profiling system which includes 10 STR regions and a sex indicating test. However, with any DNA technique, the cautious juror should not convict on genetic fingerprint evidence alone if other factors raise doubt.

Contamination with other evidence (secondary transfer) is a key source of incorrect DNA profiles and raising doubts as to whether a sample has been adulterated is a favorite defence technique. More rarely, chimerism is one such instance where the lack of a genetic match may unfairly exclude a suspect.

Evidence of Genetic Relationship

It's also possible to use DNA profiling as evidence of genetic relationship, but testing that shows no relationship isn't absolutely certain. While almost all individuals have a single and distinct set of genes, rare individuals, known as "chimeras", have at least two different sets of genes. There have been several cases of DNA profiling that falsely "proved" that a mother was unrelated to her children.

Fake DNA Evidence

The value of DNA evidence has to be seen in light of recent cases where criminals planted fake DNA samples at crime scenes. In one case, a criminal even planted fake DNA evidence in his own body: Dr. John Schneeberger raped one of his sedated patients in 1992 and left semen on her underwear. Police drew what they believed to be Schneeberger's blood and compared its DNA against the crime scene semen DNA on three occasions, never showing a match. It turned out that he had surgically inserted a Penrose drain into his arm and filled it with foreign blood and anticoagulants.

In a study conducted by the life science company Nucleix and published in the journal Forensic Science International, scientists found that an In vitro synthesized sample of DNA matching any desired genetic profile can be constructed using standard molecular biology techniques without obtaining any actual tissue from that person.

DNA Evidence as Evidence in Criminal Trials

Familial Searching

Familial searching is the use of family members' DNA to identify a closely related suspec in jurisdictions where large DNA databases exist, but no exact match has been found. The first successful use of the practice was in a UK case where a man was convicted of manslaughter when he threw a brick stained with his own blood into a moving car. Police could not get an exact match to the UK's DNA database because the man had no criminal convictions, but police implicated him using a close relative's DNA. The technique was used to catch a Los Angeles serial killer known as the "Grim Sleeper" in 2010. However, critics have challenged the technology as "deeply antithetical to American values" and likely to lead to excess investigation of racial and ethnic minorities.

Surreptitious DNA Collecting

Police forces may collect DNA samples without the suspects' knowledge, and use it as evidence. Legality of this mode of proceeding has been questioned in Australia. In the United States, it has been accepted, courts often claiming that there was no expectation of privacy, citing California v. Greenwood (1985), during which the Supreme Court held that the Fourth Amendment does not prohibit the warrantless search and seizure of garbage left for collection outside the curtilage of a home. Critics of this practice underline the fact that this analogy ignores that "most people have no idea that they risk surrendering their genetic identity to the police by, for instance, failing to destroy a used coffee cup. Moreover, even if they do realize it, there is no way to avoid abandoning one's DNA in public." In the UK, the Human Tissue Act 2004 prohibited private individuals from covertly collecting biological samples (hair, fingernails, etc.) for DNA analysis, but excluded medical and criminal investigations from the offense.

England and Wales

Evidence from an expert who has compared DNA samples must be accompanied by evidence as to the sources of the samples and the procedures for obtaining the DNA profiles. The judge must ensure that the jury must understand the significance of DNA matches and mismatches in the profiles. The judge must also ensure that the jury does not confuse the 'match probability' (the probability that a person that is chosen at random has a matching DNA profile to the sample from the scene) with the 'likelihood ratio' (the probability that a person with matching DNA committed the crime).

In *R v. Doheny*, EWCA Crim 728. Phillips LJ gave this example of a summing up, which should be carefully tailored to the particular facts in each case:

Members of the Jury, if you accept the scientific evidence called by the Crown, this indicates that there are probably only four or five white males in the United Kingdom from whom that semen stain could have come. The Defendant is one of them. If that is the position, the decision you have to reach, on all the evidence, is whether you are sure that it was the Defendant who left that stain or whether it is possible that it was one of that other small group of men who share the same DNA characteristics. Juries should weigh up conflicting and corroborative evidence, using their own common sense and not by using mathematical formulae, such as Bayes' theorem, so as to avoid "confusion, misunderstanding and misjudgment".

Presentation and Evaluation of Evidence of Partial or Incomplete DNA Profiles

R v Bates (2006) EWCA Crim 1395 Moore-Bick LJ said: "We can see no eason why partial profile DNA evidence should not be admissible provided that the jury are made aware of its nherent limitations and are given a sufficient explanation toenable them to evaluate it.

There may be cases where the match probability in relation to all the samples tested is so great that the judge would consider its probative value to be minimal and decide to exclude the evidence in the exercise of his discretion, but this gives rise to no new question of principle and can be left for decision on a case by case basis.

However, the fact that there exists in the case of all partial profile evidence the possibility that a "missing" allele might exculpate the accused altogether does not provide sufficient grounds for rejecting such evidence.

In many there is a possibility (at least in theory) hat evidenceexists whichwould assist the accused and perhaps even exculpate him altogether, but that does not provide grounds for excluding relevant evidence that i available and otherwise admissible, though it does make it important to ensure that the jury are given sufficient information to enable them to evluate that evidence properly".

DNA Testing in the US

There are state lws on DNA profiling in all 50 states of the United States. Detailed information on database laws in each state can be found at the National Conference of State Legislatures website.

Development of Artificial DNA

In August 2009, scintists in Israel stunned the forensic sciences and raised serious questions concerning the use of DNA by law enforcement as the ultimate method of identification. In apaper published inthe journal *Forensic Science International: Genetics*, the Israeli researchers demonstrated that it is possible to manufacture DNA in a laboratory, and thus fasify DNA evidence. The scientists fabricated saliva and blood samples, which originally contained DNA from a person other than the ostensible donor of the blood and saliva.

The researchers also showed that, using a DNA database, it is possible to take information from a profile and manufacture DNA to match it, and that this can be done without access to any actual DNA from the person whose DNA they are duplicating. The synthetic DNA oligos equired for the procedure are common in molecular laboratories. Dr. Daniel Frumkin, lead author on the paper, was quoted in *The New York Times* s saying, "You can just engineer a crime scene... any biology undergraduate could perform this." Dr. Frumkin perfected a test that can forensically differentiate real DNA samples from fake ones. His test detects epigenetic modifications, in particular, DNA methylation. Seventy percent of the DNA in any human genome is methylated, meaning it contains methyl group modifications within a CpG dinucleotide context. Methylation at the promoter region is associated with gene silencing. The synthetic DNA lacks this epigenetic modification, which allows the test to istinguish manufactured DNA from original, genuine, DNA.

It is unknown how many, if any, police departments currently use the test, which appears to be a serious issue in light of Frumkin's claim that the DNA manufacturing procedure is within the grasp of any undergraduate biology student. No police lab has publicly announced that it is using the new test to verify DNA results, while FSI Genetics says that any forensic laboratory doing DNA identification should adopt this test to authenticate its results as "real" DNA.

Bioinformatics

Bioinformatics involves the manipulation, searching, and data mining of DNA sequence data. The development of techniques to store and search DNA sequences have led to widely applied advances in computer science, especially string searching algorithms, machine learning and database theory. String searching or matching algorithms, which find an occurrence of a sequence of letters inside a larger sequence of letters, were developed to search for specific sequences of nucleotides. In other applications such as text editors, even simple algorithms for this problem usually suffice, but DNA sequences cause these algorithms to exhibit near-worst-case behaviour due to their small number of distinct characters.

The related problem of sequence alignment aims to identify homologous sequences and locate the specific mutations that make them distinct. These techniques, especially multiple sequence alignment, are used in studying phylogenetic relationships and protein function.

Data sets representing entire genomes' worth of DNA sequences, such as those produced by the Human Genome Project, are difficult to use without annotations, which label the locations of genes and regulatory elements on each chromosome. Regions of DNA sequence that have the characteristic patterns associated with protein- or RNA-coding genes can be identified by gene finding algorithms, which allow researchers to predict the presence of particular gene products in an organism even before they have been isolated experimentally.

DNA Nanotechnology

DNA nanotechnology uses the unique molecular recognition properties of DNA and other nucleic acids to create self-assembling branched DNA complexes with useful properties. DNA is thus used as a structural material rather than as a carrier of biological information.

This has led to the creation of two-dimensional periodic lattices (both tile-based as well as using the "DNA origami" method) as well as three-dimensional structures in the shapes of polyhedra. Nanomechanical devices and algorithmic self-assembly have also been demonstrated, and these DNA structures have been used to template the arrangement of other molecules such as gold nanoparticles and streptavidin proteins.

History and Anthropology

Because DNA collects mutations over time, which are then inherited, it contains historical information, and, by comparing DNA sequences, geneticists can infer the evolutionary history of organisms, their phylogeny. This field of phylogenetics is a powerful tool in evolutionary biology. If DNA sequences within a species are compared, population geneticists can learn the history of particular populations. This can be used in studies ranging from ecological genetics to anthropology; For example, DNA evidence is being used to try to identify the Ten Lost Tribes of Israel.

DNA has also been ued to look at modern family relationships, such as establishing family relationships between the descendants of Sally Hemings and Thomas Jefferson. This usage is closely related to the use of DNA in criminal investigations detailed above. Indeed, some criminal investigations have been solved when DNA from crime scenes has matched relatives of the guilty individual.

History of DNA Research

DNA was first isolated by the Swiss physician Friedrich Miescher who, in 1869, discovered a microscopic substance in the pus of discarded surgical bandages. As it resided in the nuclei of cells, he called it "nuclein". In 1919, Phoebus Levene identified the base, sugar and phosphate nucleotide unit. Levene suggested that DNA consisted of a string of nucleotide units linked together through the phosphate groups. However, Levene thought the chain was short and the bases repeated in a fixed order. In 1937 William Astbury produced the first X-ray diffraction patterns that showed that DNA had a regular structure. In 1928, Frederick Griffith discovered that traits of the "smooth" form of the *Pneumococcus* could be transferred to the "rough" form of the same bacteria by mixing killed "smooth" bacteria with the live "rough" form. This system provided the first clear suggestion that DNA carries genetic information—the Avery-MacLeod-McCarty experiment—when Oswald Avery, along with coworkers Colin MacLeod and Maclyn McCarty, identified DNA as the transforming principle in 1943. DNA's role in heredity was confirmed in 1952, when Alfred Hershey and Martha Chase in the Hershey-Chase experiment showed that DNA is the genetic material of the T2 phage.

In 1953, James D. Watson and Francis Crick suggested what is now accepted as the first correct double-helix model of DNA structure in the journal *Nature*. Their double-helix, molecular model of DNA was then based on a single X-ray diffraction image (labeled as "Photo 51") taken by Rosalind Franklin and Raymond Gosling in May 1952, as well as the information that the DNA bases are paired — also obtained through private communications from Erwin Chargaff in the previous years. Chargaff's rules played a very important role in establishing double-helix configurations for B-DNA as well as A-DNA. Experimental evidence supporting the Watson and Crick model were published in a series of five articles in the same issue of *Nature*.

Of these, Franklin and Gosling's paper was the first publication of their own X-ray diffraction data and original analysis method that partially supported the Watson and Crick model; this issue also contained an article on DNA structure by Maurice Wilkins and two of his colleagues, whose analysis and *in vivo* B-DNA X-ray patterns also supported the presence *in vivo* of the double-helical DNA configurations as proposed by Crick and Watson for their double-helix molecular model of DNA in the previous two pages of *Nature*. In 1962, after Franklin's death, Watson, Crick, and Wilkins jointly received the Nobel Prize in Physiology or Medicine. However, Nobel rules of the time allowed only living recipients, but a vigorous debate continues on who should receive credit for the discovery.

In an influential presentation in 1957, Crick laid out the "Central Dogma" of molecular biology, which foretold the relationship between DNA, RNA, and proteins, and articulated the "adaptor hypothesis". Final confirmation of the replication mechanism that was implied by the double-helical structure followed in 1958 through the Meselson-Stahl experiment. Further work by Crick and coworkers showed that the genetic code was based on non-overlappin triplets of bases, called codons, allowing Har Gobind Khorana, Robert W. Holley and Marshall Warren Nirenberg to decipher the genetic code. These findings represent the birth of molecular biology.

Protein Synthesis

Protein synthesis is the process in which cells build proteins. The term is sometimes used to refer only to protein translation but more often it refers to a multi-step process, beginning with amino acid synthesis and transcription of nuclear DNA into messenger RNA, which is then used as input to translation.

The cistron DNA is transcribed into a variety of RNA intermediates. The last version is used as a template in synthesis of a polypeptide chain. Proteins can often be synthesized directly from genes by translating mRNA. When a protein needs to be available on short notice or in large quantities, a protein precursor is produced. A proprotein is an inactive protein containing one or more inhibitory peptides that can be activated when the inhibitory sequence is removed by proteolysis during posttranslational modification. A preprotein is a form that contains a signal sequence (an N-terminal signal peptide) that specifies its insertion into or through membranes; i.e., targets them for secretion. The signal peptide is cleaved off in the endoplasmic reticulum. Preproproteins have both sequences (inhibitory and signal) still present.

For synthesis of protein, a succession of tRNA molecules charged with appropriate amino acids have to be brought together with an mRNA molecule and matched up by base-pairing through their anti-codons with each of its successive codons. The amino acids then have to be linked together to extend the growing protein chain, and the tRNAs, relieved of their burdens, have to be released. This whole complex of processes is carried out by a giant multimolecular machine, the ribosome, formed of two main chains of RNA, called ribosomal RNA (rRNA), and more than 50 different proteins.

This molecular juggernaut latches onto the end of an mRNA molecule and then trundles along it, capturing loaded tRNA molecules and stitching together the amino acids they carry to form a new protein chain. Protein biosynthesis, although very similar, is different for prokaryotes and eukaryotes.

Amino Acid Synthesis

Amino acid synthesis is the set of biochemical processes (metabolic pathways) by which the various amino acids are produced from other compounds. The substrates for these processes are various compounds in the organism's diet or growth media. Not all organisms are able to synthesise all amino acids, for example humans are only able to synthesise 12 of the 20 standard amino acids.

A fundamental problem for biological systems is to obtain nitrogen in an easily usable form. This problem is solved by certain microorganisms capable of reducing the inert Na"N molecule (nitrogen gas) to two molecules of ammonia in one of the most remarkable reactions in biochemistry. Nitrogen in the form of ammonia is the source of nitrogen for all the amino acids. The carbon backbones come from the glycolytic pathway, the pentose phosphate pathway, or the citric acid cycle.

In amino acid production, one encounters an important problem in biosynthesis— namely, stereochemical control. Because all amino acids except glycine are chiral, biosynthetic pathways must generate the correct isomer with high fidelity.

In each of the 19 pathways for the generation of chiral amino acids, the stereochemistry at the á-carbon atom is established by a transamination reaction that involves pyridoxal phosphate. Almost all the transaminases that catalyze these reactions descend from a common ancestor, illustrating once again that effective solutions to biochemical problems are retained throughout evolution.

Biosynthetic pathways are often highly regulated such that building blocks are synthesized only when supplies are low. Very often, a high concentration of the final product of a pathway inhibits the activity of enzymes that function early in the pathway. Often present are allosteric enzymes capable of sensing and responding to concentrations of regulatory species. These enzymes are similar in functional properties to aspartate transcarbamoylase and its regulators. Feedback and allosteric mechanisms ensure that all twenty amino acids are maintained in sufficient amounts for protein synthesis and other processes.

Amino acids are synthesized from α-ketoacids, and later transaminated from another aminoacid, usually Glutamate. The enzyme involved in this reaction is an aminotransferase.

Glutamate itself is formed by amination of α-ketoglutarate:

Nitrogen Fixation: Microorganisms Use ATP and a Powerful Reductant to Reduce Atmospheric Nitrogen to Ammonia

Microorganisms use ATP and reduced ferredoxin, a powerful reductant, to reduce N2 to NH3. An iron-molybdenum cluster in nitrogenase deftly catalyzes the fixation of N2, a very inert molecule. Higher organisms consume the fixed nitrogen to synthesize amino acids, nucleotides, and other nitrogen-containing biomolecules. The major points of entry of NH4+ into metabolism are glutamine or glutamate.

Amino Acids are Made from Intermediates of the Citric Acid Cycle and Other Major Pathways

Of the basic set of 20 amino acids (not counting selenocysteine) there are 8 that human beings cannot synthesize. In addition, the amino acids arginine, cysteine, glycine, glutamine, histidine, proline, serine and tyrosine are considered conditionally essential, meaning they are not normally required in the diet, but must be supplied exogenously to specific populations that do not synthesize it in adequate amounts. (For example, enough arginine is synthesized by the urea cycle to meet the needs of an adult but perhaps not those of a growing child.) Amino acids that need to be obtained from the diet are called essential amino acids. Nonessential amino acids are produced in the body.

The pathways for the synthesis of nonessential amino acids are quite simple. Glutamate dehydrogenase catalyzes the reductive amination of α-ketoglutarate to glutamate. A transamination reaction takes place in the synthesis of most amino acids. At this step, the chirality of the amino acid is established. Alanine and aspartate are synthesized by the transamination of pyruvate and oxaloacetate, respectively. Glutamine is synthesized from NH4+ and glutamate, and asparagine is synthesized similarly. Proline and arginine are derived from glutamate.

Serine, formed from 3-phosphoglycerate, is the precursor of glycine and cysteine. Tyrosine is synthesized by the hydroxylation of phenylalanine, an essential amino acid. The pathways for the biosynthesis of essential amino acids are much more complex than those for the nonessential ones.

Tetrahydrofolate, a carrier of activated one-carbon units, plays an important role in the metabolism of amino acids and nucleotides. This coenzyme carries one-carbon units at three oxidation states, which are interconvertible: most reduced—methyl; intermediate—methylene; and most oxidized—formyl, formimino, and methenyl. The major donor of activated methyl groups is S-adenosylmethionine, which is synthesized by the transfer of an adenosyl group from ATP to the sulfur atom of methionine. S-Adenosylhomocysteine is formed when the activated methyl group is transferred to an acceptor. It is hydrolyzed to adenosine and homocysteine, the latter of which is then methylated to methionine to complete the activated methyl cycle. The A Cortisol inhibites protein synthesis.

Amino Acid Biosynthesis is Regulated by Feedback Inhibition

Most of the pathways of amino acid biosynthesis are regulated by feedback inhibition, in which the committed step is allosterically inhibited by the final product. Branched pathways require extensive interaction among the branches that includes both negative and positive regulation. The regulation of glutamine synthetase from E. coli is a striking demonstration of cumulative feedback inhibition and of control by a cascade of reversible covalent modifications.

Amino Acids are Precursors of Many Biomolecules

Amino acids are precursors of a variety of biomolecules. Glutathione serves as a sulfhydryl buffer and detoxifying agent. Glutathione peroxidase, a selenoenzyme, catalyzes the reduction of hydrogen peroxide and organic peroxides by glutathione. Nitric oxide, a short-lived messenger, is formed from arginine. Porphyrins are synthesized from glycine and succinyl CoA, which condense to give β-aminolevulinate. Two molecules of this intermediate become linked to form porphobilinogen. Four molecules of porphobilinogen combine to form a linear tetrapyrrole, which cyclizes to uroporphyrinogen III. Oxidation and side-chain modifications lead to the synthesis of protoporphyrin IX, which acquires an iron atom to form heme.

Transcription

In transcription an mRNA chain is generated, with one strand of the DNA double helix in the genome as template. This strand is called the template strand. Transcription can be divided into 3 stages: Initiation, Elongation and Termination, each regulated by a large number of proteins such as transcription factors and coactivators that ensure the correct gene is transcribed.

The DNA strand is read in the 3' to 5' direction and the mRNA is transcribed in the 5' to 3' direction by the RNA polymerase.

Transcription occurs in the cell nucleus, where the DNA is held. The DNA structure of the cell is made up of two helixes made up of sugar and phosphate held together by the bases. The sugar and the phosphate are joined together by covalent bond. The DNA is "unzipped" by the enzyme helicase, leaving the single nucleotide chain open to be copied. RNA polymerase reads the DNA strand from 3 prime (3') end to the 5 prime (5') end, while it synthesizes a single strand of messenger RNA in the 5' to 3' direction.

The general RNA structure is very similar to the DNA structure, but in RNA the nucleotide uracil takes the place that thymine occupies in DNA. The single strand of mRNA leaves the nucleus through nuclear pores, and migrates into the cytoplasm. The first product of transcription differs in prokaryotic cells from that of eukaryotic cells, as in prokaryotic cells the product is mRNA, which needs no post-transcriptional modification, while in eukaryotic cells, the first product is called primary transcript, that needs post-transcriptional modification (capping with

7 methyl guanosine, tailing with a poly A tail) to give hnRNA (heterophil nuclear RNA). hnRNA then undergoes splicing of introns (non coding parts of the gene) via spliceosomes to produce the final mRNA.

Translation

The synthesis of proteins is known as translation. Translation occurs in the cytoplasm, where the ribosomes are located. Ribosomes are made of a small and large subunit that surround the mRNA. In translation, messenger RNA (mRNA) is decoded to produce a specific polypeptide according to the rules specified by the trinucleotide genetic code. This uses an mRNA sequence as a template to guide the synthesis of a chain of amino acids that form a protein. Translation proceeds in four phases: activation, initiation, elongation, and termination (all describing the growth of the amino acid chain, or polypeptide that is the product of translation).

In activation, the correct amino acid (AA) is joined to the correct transfer RNA (tRNA). While this is not technically a step in translation, it is required for translation to proceed. The AA is joined by its carboxyl group to the 3' OH of the tRNA by an ester bond. When the tRNA has an amino acid linked to it, it is termed "charged". Initiation involves the small subunit of the ribosome binding to 5' end of mRNA with the help of initiation factors (IF), other proteins that assist the process. Elongation occurs when the next aminoacyl-tRNA (charged tRNA) in line binds to the ribosome along with GTP and an elongation factor.

Termination of the polypeptide happens when the A site of the ribosome faces a stop codon (UAA, UAG, or UGA). When this happens, no tRNA can recognize it, but releasing factor can recognize nonsense codons and causes the release of the polypeptide chain. The capacity of disabling or inhibiting translation in protein biosynthesis is used by antibiotics such as: anisomycin, cycloheximide, chloramphenicol, tetracycline, streptomycin, erythromycin, puromycin etc.

Translation is the process of converting the mRNA codon sequences into an amino acid polypeptide chain.

1. Amino acid activation
2. Initiation - A ribosome attaches to the mRNA and starts to code at the FMet codon (usually AUG, sometimes GUG or UUG).
3. Elongation - tRNA brings the corresponding amino acid (which has an anticodon that identifies the amino acid as the corresponding molecule to a codon) to each codon as the ribosome moves down the mRNA strand.

4. Termination - Reading of the final mRNA codon (aka the STOP codon), which ends the synthesis of the peptide chain and releases it.

Events Following Protein Translation

The events following biosynthesis include post-translational modification and protein folding. During and after synthesis, polypeptide chains often fold to assume, so called, native secondary and tertiary structures. This is known as *protein folding*.

Many proteins undergo *post-translational modification*. This may include the formation of disulfide bridges or attachment of any of a number of biochemical functional groups, such as acetate, phosphate, various lipids and carbohydrates. Enzymes may also remove one or more amino acids from the leading (amino) end of the polypeptide chain, leaving a protein consisting of two polypeptide chains connected by disulfide bonds.

In general, protein molecules are believed to be modified by small chemical groups, post-translationally. Chemical modifications such as, phosphorylation of serine/threonine, acetylation or methylation of lysine, hydroxylation of proline/lysine, formylation of glycine, glycosylation of serine/threonine/asparagine, acylation of cysteine, myristoylation of glycine, biotinylation of lysine, ubiquitination, etc. on proteins is a very important issue in relation to properly understanding the biological functions of a given protein.

These much-studied post-translational modifications have become well-established with the discovery of the respective enzymes (kinases for phosphorylation, acetylases for acetylation, methyl transferases for methylation, etc.), which carry on the chemical modifications on the specific amino acid residues. All of these modifications are still believed to have happened as post-translational events. There is no study yet on when actually one particular modification occurs on a given amino acid residue in a given protein. Does it happen when the protein is already formed, or when the amino acid chain is being synthesized, or before the translation of the primary chain has begun?

Since these chemical modifications are related to the biological functions of a protein, it is easy to think that these chemical modifications have happened to the whole protein molecule, after the protein primary chain is fully synthesized; but, if that is the case, we have to consider the fact that the primary chains get folded instantly, (in a similar way as the newly synthesized DNA strands form helixes), to attain its compact-globular conformation ; As most of the primary chains are fairly long [a 5Kd protein may have 40-45 amino acid residues in its primary chain], it is likely that the newly formed amino acid chain tries to remain intact by folding, thereby avoiding its breakdown via lots of proteases present within the cytoplasm.

And, no capping event to protect the N-terminal end of the primary sequence (similar to 5' m-RNA capping to protect m-RNAs) is ever discovered for protein primary structure. So, by folding mechanism, the primary chain, perhaps, avoids the protease attacks.

However, once it gets folded, it may be very difficult for the respective enzyme molecule to find out the particular aa residue from the complexity of that compactly folded conformation. In addition, it can be clearly imagined that this enzymatic modification/reaction on a given amino acid requires presence and association of the appropriate ezyme, necessary cofactors, etc.

This association is much easier to occur when the amino acid residues in the primary structure are readily avaiable for binding; in othe words, it is much difficult for the enzyme molecules to find and to bind to its substrate amino acid residue in a mature protein molecule after its three-dimensional conformation been attained.

So one can think that the modifications can happen while the primary chain of the protein is being synthesized during the translation process on the m-RNA strand; the amino acid residues on the primary chain can be modified instantly and enzymatically by kinases, acetylases, hydroxylases, methyltransferases, etc. to initiate proper folding for protection in order to avoid degradation by proteases, thereby gaining the globular form.

Also, the reader can imagine another scenario in which, while the free amino acid molecules are formed within a cell and become available, they [in that free state] may be modified enzymatically before taking the ride to the translational event; this means that, while the primary chain is being synthesized, the pre-modified amino acid molecules are ready to be engaged.

This way, as the primary chain is getting synthesized, it does not have to be modified by the modifying enzymes anymore, and can fold itself instantaneously without concern of degradation. This also arises new thoughts that

(i) phosphate, acetyl, methyl, biotinyl, acyl, etc. groups may have the ability to inhibit protease actions on the primary chains;

(ii) most or all of the amino acid residues get modified by small chemical groups (so far, only some chemical groups are known).

It is still not fully known exactly when and how the actual modification of a given amino acid residue occurs, at which stage of synthesis, within a protein molecule. Once the chemically-modified, protease-insensitive, intact protein molecule is generated, it must perform its biological function, which requires its being activated. The activation is probably done by a second set of enzymatic reactions when these chemical groups are removed from the a residues (or added back). So, the second type of post-translational modifications are the opposite reactions of the above-described type, which are dephosphorylation by phosphatases or phosphoryl transferases, deacetylation by deacetylases or acetyltransferases, demethylation by demethylases or methyltransferases, ubiquitination, SuMoylation, glycosylation, biotinylation, etc. These are taking place on the whole protein molecule toward generating its activated form to execute a particular function or toward its deactivation followed by its total degradation.

Chapter 8

Techniques in Molecular Biology

Molecular biology is the branch of science which deals with the structure and function of macromolecules. It has various techniques which are also applied in other branches of biology.

Here are several techniques used in molecular biology which are also related to fields like genetics, recombinant DNA technology etc. These techniques include following:

Expression cloning - One of the most widely used techniques of molecular biology to study protein function is expression cloning. In this technique DNA coding for a protein of interest is cloned. This is done by using PCR and/or restriction enzymes.

The DNA encoding the required protein is inserted into a plasmid which is known as an expression vector. This plasmid may include special promoter elements to drive production of the protein of interest, and may also have antibiotic resistance markers to help follow the plasmid.

Polymerase Chain Reaction - is most frequently uses technique for copying DNA into millions of clones within few minutes. It can be used to insert restriction enzymes sites in a DNA fragment or to mutate a particular base sequence and also for determination about the presence of a particular DNA fragment in cDNA libraries.

Gel electrophoresis - is used to separate fragments of DNA, RNA and proteins using different properties like according to size, ionic bond etc. For example agarose gel is used to separate RNA and DNA according to their size and SDS-PAGE is used to separate different proteins.

Blotting techniques - is the technique of probing DNA(Southern blotting), RNA(Northern blotting), or proteins(Western blotting) molecules. Eastern blotting is used to recognize post translational modifications of proteins.

Other well known and essential techniques in molecular biology include arrays like DNA microarray, allele specific oligonucleotide(ASO) etc.

Role of Celular Biology

Main tool used in cellular biology is microscope. A wide range of microscopes including Light microscope, Electron microscope, Fluorescent microscope are used for this purpose.

The main purpose of cellular biology is to study different cellular organelles present in all types(prokaryotic-eukaryotic, plant-animal etc) of cells. These organelles include- nucleus, mitochondria, chloroplast, cytoplasm, endoplasmic reticulam, golgi apparatus, cell wall. cell membrane and many more cellular organelles.

Techniques involved in cellular biology other than microscopy include following:

1. Cell culture - in vitro technique of growing cells in laboratory.
2. Immunostaining - is the technique used to locate proteins present in the cell or parts of tissues.
3. Transfection - involve introduction of new gene into a cell
4. Cell fractionation - used to obtain cellular organelles for their study.

Other techniques used in cellular biology include centrifugation, flow cytometry, immunoprecipitation, gene knockdown etc.

Polymerase Chain Reaction

Polymerase chain reaction is a technique of molecular biology which makes multiple copies of DNA from a single fragment. The process of polymerase chain reaction takes place as follows:

1. First the DNA molecule is denatured by heating under the temperature of 94 to 98 degree centigrade.
2. When the strands are separated, they are bound with DNA primers with the temperature 50 to 65 degree Celsius.
3. After annealing, the elongation of the separated DNA strands take place by using heat stable DNA polymerases like Taq (Thermus aquaticus).

There is another method of PCR named reverse transcription PCR which is used to amplify the RNA molecules, while real time PCR is used to measure the DNA o RNA molecule quantitatively.

Expression Cloning

Cloning is the most useful technique of molecular biology because it is used to study the function of the proteins. This technique works in the way that it clones the DNA coding which makes protein, using the process of polymerase chain reaction or restriction enzymes and made multiple copies of that coding. The cloning converts the DNA samples into plasmid vectors. The plasmid either has promoter elements which make possible the production of protein of interest or they have the antibiotic resistant markers.

The plasmid produced from cloning can be inserted in the bacterial cell or the animal cell depending on the mode of experiment. If it is inserted in the bacterial cell, then three processes are involved which make possible the insertion of plasmid into the bacterial cell. These processes are transformation, conjugation and transduction. When the insertion of the cloned plasmid in the animal cell is concerned then the process of transfection is involved. Transfection techniques are of various types for example, microinjection, calcium phosphate transfection and liposome transfection.

But it is also possible that the plasmid can be inserted in the animal cell by bacteria or virus. If bacteria are involved for insertion then it is called as bactofention and mainly the specie of Agrobacterium tumafeciens is used for this purpose.

Plasmid can stay in the eukaryotic cell in two forms; either it mixes up with the genome of the host cell which results in the stable transfection or it stays independently in the cell by transient transfection. In both the cases, the protein of interest is in the cell and can be expressed. If the cell contains specific cell signaling factors or inducible promoters then protein is expressed efficiently than without these factors.

Gel Electrophoresis

Gel electrophoresis is the basic technique of molecular biology which is used to separate DNA, RNA or protein molecules. The method involved is that, the molecules are placed in the immobilized gel and an electric current is passed through it. This current separates the molecules according to the size of the strands. Similarly proteins are also separated in the gel by using an SDS-PAGE gel or it can also be separated according to their size and the charge which they carry.

Allele Specific Oligonucleotide

Allele specific technique is used to detect the single base mutations in the genes. It does not need any gel electrophoresis or polymerase chain reaction. In this technique, short probes are used usually 20 to 25 nucleotides long and they are also labeled.

The labeled probes are exposed to the non-fragment target DNA; as a result hybridization takes place due to the short length probes. If there is only a single change in the nucleotide base, then the hybridization will stop. After this step, the target DNA is washed and those probes are removed which did not hybridize. Either radiation or florescence detects the presence of probe on the DNA fragment.

Arrays

This is the technique of molecular biology which uses DNA arrays in the form of spots which are attached to the solid support. This solid support is usually microscopic slide which contains the single stranded DNA oligonucleotide fragment present on each spot.

This technique is very feasible because it contains large amount of very short spots with DNA fragments on each slide. The spot is so small that it contains a single DNA fragment. This fragment is complementary to a single DNA sequence. The process starts with the extraction of RNA molecule from a tissue which is converted in the form of cDNA. The DNA fragments present on the spots in the slide are then hybridized with these cDNAs. They can be visualized after hybridization.

This technique is useful in the gene expression data as many arrays can be made with the same position of the DNA fragments on the spots, so that the gene expression of healthy and diseases genes can be compared. For example common yeast consists of 7000 genes. With the microarray, it is possible to observe the expression of each gene qualitatively. It is also possible to view the changes in the expression of these genes due to temperature.

There are also other methods of construction of arrays instead of suing DNA fragments. For example an antibody array can be used to determine the presence f proteins or bacteria in the sample of blood.

Protein Folding and Modifications

A polypeptide chain is synthesized on large cellular structures, the ribosomes, by a complex process in which assembly of amino acids in a particular sequence is dictated by messenger RNA (mRNA). The nascent polypeptide chain undergoes folding and, in many cases, chemical modification to generate the final protein.

Any polypeptide chain containing *n* residues could, in principle, fold into 8^n conformations. This value is based on the fact that only eight bond angles are stereochemically allowed in the polypeptide backbone. In general, however, all molecules of any protein species adopt a single conformation, called the *native state,* which is the most stably folded form of the molecule. Misfolding to non-native conformations is suppressed by two distinct mechanisms. At the molecular level, a protein folds through a pathway that favors only a few intermediate steps. Furthermore, a cellular system prevents misfolded proteins from forming. After a protein has carried out its functions, specific sequences that limit the life span of the protein target it for degradation.

The Information for Protein Folding is Encoded in the Sequence

The realization that the amino acid sequence of a protein determines its folding came from in vitro studies on protein unfolding and refolding. Thermal energy from heat, extremes of pH that alter the charges on amino acid side chains, and chemicals such as urea or guanidine hydrochloride at concentrations of 6–8 M can disrupt the weak noncovalent bonds that stabilize the native conformation of a protein. The denaturation resulting from such treatment causes a protein to lose both its compact conformation and activity.

Most denatured proteins precipitate in solution because hydrophobic groups, normally buried inside the molecules, interact with similar regions of other unfolded molecules, causing them to form an insoluble aggregate.

Many proteins that are completely unfolded in 8 M urea and β-mercaptoethanol (which reduces disulfide bonds) can *renature* (refold) into their native state when the denaturing reagents are removed by dialysis. During renaturation, all the disulfide, hydrogen, and hydrophobic bonds that stabilize the native conformation are re-formed. Thus, in this case proteins can be carried through a denaturationrenaturation cycle, which first destroys and then reestablishes their original structure and function. Because renaturation requires no cofactors or other proteins, at least in the test tube, protein folding is a self-assembly process.

Treatment with an 8 M urea solution containing mercaptoethanol ($HSCH_2CH_2OH$) completely denatures most proteins. The urea breaks intramolecular hydrogen and hydrophobic bonds, and the mercaptoethanol reduces each disulfide bridge (–S–S–) to two sulfhydryl (–SH) groups. When these chemicals are removed by dialysis, the –SH groups on the unfolded chain oxidize spontaneously to reform disulfide bridges, and the polypeptide chain simultaneously refolds into its native conformation.

The observation by Christian Anfinsen of such reversible denaturation and renaturation of ribonuclease, an enzyme that degrades RNA, provided a clue that the information for folding a protein lies in its sequence. A general mechanism by which proteins refold in vitro has been elucidated in experiments in which the renaturing conditions are carefully adjusted and the refolding reaction is interrupted at various time intervals. Such studies have shown that the polypeptide goes through several transient reconfigurations, including a "molten globule" state, before the native tertiary conformation is reached. In the case of ribonuclease, which has several internal disulfide bonds, the folding pathway involves rearrangements of disulfide bond pairs to the native conformation.

Under appropriate refolding conditions, the molecule condenses around a hydrophobic core into a compact, but non-native, intermediate, called a *molten globule.*In this folding intermediate, much of the secondary structure is present. Long-range interactions then form the tertiary structure, folding the molecule into its native three-dimensional conformation.

Folding of Proteins in Vivo is Promoted by Chaperones

Folding of proteins in vitro is an inefficient process, with only a minority of unfolded molecules undergoing complete folding within a few minutes. Clearly, in vivo most protein molecules must rapidly fold into their correct shape; otherwise, cells would waste much energy in the synthesis of nonfunctional proteins and in the degradation of misfolded or unfolded proteins. More than 95 percent of the proteins present within cells have been shown to be in their native conformation, despite high protein concentrations (H–100 mg/ml), which usually cause proteins to precipitate in vitro.

The explanation for the cell's remarkable efficiency in promoting protein folding probably lies in chaperones, a family of proteins found in all organisms from bacteria to humans. Chaperones are located in every cellular compartment, bind a wide range of proteins, and may be part of a general protein-folding mechanism. There are two general families of chaperones: *molecular chaperones,* which bind and stabilize unfolded or partially folded proteins, thereby preventing these proteins from being degraded; and *chaperonins,* which directly facilitate their folding.

Chaperones have ATPase activity, and their ability to bind and stabilize their target proteins is specific and dependent on ATP hydrolysis. Binding of chaperones to partially folded proteins suggests that the folding process could be regulated at intermediate steps. Molecular chaperones consist of the Hsp70 family of proteins, which includes Hsp70 in the cytosol and mitochondrial matrix, Bip in the endoplasmic reticulum, and DnaK, a bacterial chaperone. First identified by its rapid appearance after a cell has been stressed by heat shock, Hsp70 is the major chaperone protein in all organisms. When bound to ATP, Hsp70 assumes an open form in which an exposed hydrophobic pocket transiently binds to exposed hydrophobic regions of the unfolded target protein. Hydrolysis of the bound ATP causes Hsp70 to assume a closed form, releasing the target protein. Molecular chaperones are thought to bind all nascent polypeptide chains as they are being synthesized on ribosomes. In bacteria, 85 percent of the proteins are released from their chaperone and go on to fold normally; an even higher percentage of proteins in eukaryotes follow this pathway.

(a) Many proteins 1 fold into their proper three-dimensional structure with the assistance of Hsp70, a molecular chaperone that transiently binds to a nascent polypeptide as it emerges from a ribosome. Proper folding of some proteins 2 also depends on the chaperonin TCiP, a large barrel-shaped complex of Hsp60 units.

(b) GroEL, the bacterial homolog of TCiP, is a barrel-shaped complex of 14 identical 60,000-MW subunits arranged in two stacked rings. In the absence of ATP or presence of ADP, GroEL exists in a "tight" conformational state that binds partially folded or misfolded proteins. Binding of ATP shifts GroEL to a more open, "relaxed" state, which releases the folded protein.

Proper folding of a small proportion of proteins (e.g., the cytoskeletal proteins actin and tubulin) requires additional assistance, which is provided by chaperonins. Eukaryotic chaperonins, called *TCiP,* are large, barrel-shaped, multimeric complexes composed of eight Hsp60 units.

The bacterial homolog, known as *GroEL,* contains 14 identical subunits. The GroEL folding mechanism, which is better understood than TCiP-mediated folding, serves as a reasonable general model. In bacteria, a partially folded or misfolded polypeptide is inserted into the cavity of GroEL, where it binds to the inner wall and folds into its native conformation. In an ATP-dependent step, GroEL expands and the protein exits GroEL, a process assisted by a co-chaperonin, GroES, which caps the ends of GroEL. Because the eukaryotic chaperonin TCiP lacks a GroES-type co-chaperonin, the last step must differ in eukaryotes. Moreover, the size of the cavity in TCiP limits this folding pathway to polypeptides smaller than 55 kDa.

Chemical Modifications and Processing Alter the Biological Activity of Proteins

Nearly every protein in a cell is chemically altered after its synthesis on a ribosome. Such modifications may alter the activity, life span, or cellular location of proteins, depending on the nature of the alteration. Protein alterations fall into two categories: chemical modification and processing. *Chemical modification* involves the linkage of a chemical group to the terminal amino or carboxyl groups or to reactive groups in the side chains of internal residues; in some cases, these modifications are reversible.

Processing involves the removal of peptide segments and generally is irreversible. *Acetylation,* the addition of an acetyl group (CH_3CO) to the amino group of the N-terminal residue is the most common form of chemical modification, involving an estimated 80 percent of all proteins:

This modification may play an important role in controlling the life span of proteins within cells, as nonacetylated proteins are rapidly degraded by intracellular proteases. As discussed later, residues at or near the termini of some membrane proteins are chemically modified by addition of long lipidlike groups. Attachment of these hydrophobic "tails," which function to anchor proteins to the lipid bilayer, constitute one way that cells restrict certain proteins to membranes. The internal residues in proteins can be modified by attachment of a variety of chemical groups to their side chains. The most important modification is *phosphorylation* of serine, threonine, and tyrosine residues. We will encounter numerous examples of proteins whose activity is regulated by reversible phosphorylation and dephosphorylation.

The side chains of asparagine, serine, and threonine are sites for *glycosylation,* the attachment of linear and branched carbohydrate chains. Many secreted proteins and membrane proteins contain glycosylated residues. Various less common modifications are found in a limited number of proteins. Examples of modified internal residues produced by hydroxylation, methylation, and carboxylation. These modifications occur after synthesis of the polypeptide chain.

These modifications occur after synthesis of the polypeptide chain. Unlike chemical modification of residues, which often is reversible, processing of some proteins causes irreversible changes that alter their activity. In the most common form of processing, residues are removed from the C- or N-terminus of a polypeptide by cleavage of the peptide bond in a reaction catalyzed by proteases.

Proteolytic cleavage is a common mechanism of activation or inactivation, especially of enzymes involved in blood coagulation or digestion. As discussed later, the activity of certain digestive enzymes is controlled by this mechanism. Proteolysis also generates active peptide hormones, such as EGF mentioned earlier and insulin, from larger precursor polypeptides.

An unusual type of processing, termed *protein selfsplicing,* occurs in bacteria and primitive eukaryotes. *Splicing* refers to a process analogous to editing film: an internal segment of polypeptide, an intein, is removed and the ends of the polypeptide are rejoined. Unlike proteolytic processing, protein self-splicing is an autocatalytic process, which proceeds by itself without the involvement of enzymes.

The excised peptide appears to eliminate itself from the protein by a mechanism similar to that used in processing of RNA molecules. In vertebrate cells, processing of some proteins involves self-cleavage, but the subsequent ligation step is absent. One such protein is Hedgehog, which is critical to a number of developmental processes.

A segment (red) of a polypeptide, called an *intein,* is removed by cleavage of two peptide bonds (red arrows), leaving two segments (blue and green). One new peptide bond then forms between the two segments (blue arrow), regenerating a continuous polypeptide backbone. The process is autocatalytic and does not depend on enzymes. Consensus splice sites in the polypeptide chain mark the points of breakage and re-formation of the backbone. The excised segment is thought to be exposed at the surface of the folded protein.

Cells Degrade Proteins Via Several Pathways

Cells have both extracellular and intracellular pathways for degrading proteins. The major extracellular pathway is the system of *digestive proteases,* which break down ingested proteins to polypeptides in the intestinal tract. These include *endoproteases* such as trypsin and chymotrypsin, which cleave the protein backbone adjacent to basic and aromatic residues; *exopeptidases,* which sequentially remove residues from the N-terminus (aminopeptidases) or C-terminus (carboxypeptidases) of proteins; and *peptidases,* which split oligopeptides into di- and tripeptides and individual amino acids. These small molecules then are transported across the intestinal lining into the bloodstream.

The life span of intracellular proteins varies from as short as a few minutes for mitotic cyclins, which help regulate passage through mitosis, to as long as the age of an organism for proteins in the lens of the eye. Cells have several intracellular proteolytic pathways for degrading misfolded or denatured proteins, normal proteins whose concentration must be decreased, and foreign proteins taken up by the cell. One major intracellular pathway involves degradation by enzymes within lysosomes, membrane-limited organelles whose interior is acidic. Distinct from the lysosomal pathway are cytosolic mechanisms for degrading proteins.

The best-understood pathway, the ubiquitin-mediated pathway, involves two steps: addition of a chain of ubiquitin molecules to an internal lysine side chain of a target protein and proteolysis of the ubiquitinated protein by a proteasome, a large, cylindrical multisubunit complex. The numerous proteasomes present in the cell cytosol proteolytically cleave ubiquitin-tagged proteins in an ATP-dependent process that yields peptides and intact ubiquitin molecules.

A conjugating enzyme catalyzes formation of a peptide bond between ubiquitin (Ub) and the side-chain $-NH_2$ of a lysine residue in a target protein. Additional Ub molecules are added, forming a multiubiquitin chain. This chain is thought to direct the tagged protein to a proteasome, which cleaves the protein into numerous small peptide fragments. (b) Computer-generated image reveals cylindrical structure of a proteasome with a cap at each end of a core. Proteolysis of ubiquitin-tagged proteins occurs along the inner wall of the core.

To be targeted for degradation by the ubiquitinmediated pathway, a protein must contain a structure that is recognized by a ubiquitinating enzyme complex. Different conjugating enzymes recognize different degradation signals in target proteins. For example, the internal sequence Arg-X-X-Leu-Gly-X-Ile-Gly-Asx in mitotic cyclin is recognized by the ubiquitin-conjugating enzyme E1. Internal sequences enriched in proline, glutamic acid, serine, and threonine (PEST sequences) are recognized by other enzymes. The life span of many cytosolic proteins is correlated with the identity of the N-terminal residue, suggesting that certain residues at the N-terminus favor rapid ubiquitination. For example, short-lived proteins that are degraded within 3 minutes in vivo commonly have Arg, Lys, Phe, Leu, or Trp at their N-terminus. In contrast, a stabilizing amino acid such as Cys, Ala, Ser, Thr, Gly, Val, or Met is present at the N-terminus in long-lived proteins that resist proteolytic attack for more than 30 hours. All newly synthesized proteins have methionine, a stabilizing amino acid, at the N-terminus.

Thus subsequent enzymatic alteration that generates one of the destabilizing amino acids at the N-terminus is necessary to target a protein for degradation.

Aberrantly Folded Proteins are Implicated in Slowly Developing Diseases

As noted earlier, each protein species normally folds into a single, energetically favorable conformation that is specified by its amino acid sequence. Recent evidence suggests, however, that a protein may fold into an alternative three- dimensional structure for reasons that have not yet been identified. Such "misfolding" not only leads to a loss of the normal function of a protein but also marks it for proteolytic degradation. The subsequent accumulation of proteolytic fragments contributes to certain degenerative diseases characterized by the presence of insoluble protein plaques in various organs including the liver and brain.

Alzheimer's disease, for example, is marked by formation of plaques and tangles in a deteriorating brain. The filaments composing these structures are derived from proteolytic products of abundant natural proteins such as amyloid precursor protein, a transmembrane protein, and Tau, a microtubule-binding protein. Plaques in other organs are formed from proteolytic fragments of natural proteins such as gelsolin, an actin-binding protein, and serum albumin, a blood protein.

The polypeptide fragments liberated by proteolysis polymerize into very stable filaments. A degeneration of the brain, similar to that seen in Alzheimer's disease, is thought to be caused by *prions,* an infectious protein agent derived by proteolysis and re-folding of a normal brain protein. The amyloid plaque in the brain of an Alzheimer's patient appears as a tangle of filaments. (b) In the atomic force microscope, the filaments are seen to be regular arrangements of a short 47-residue fragment, called β-*amyloid peptide,* produced by proteolysis of amyloid precursor protein.

- The amino acid sequence of a protein dictates its folding into a specific three-dimensional conformation, the native state.
- Folding of denatured proteins in vitro proceeds through intermediates having secondary and non-native tertiary structure.
- Protein folding in vivo occurs with the assistance of two types of special proteins. Molecular chaperones (Hsp70 proteins) bind to nascent polypeptides emerging from ribosomes and prevent

their misfolding. Chaperonins, large complexes of Hsp60-like proteins, shelter some partially folded or misfolded proteins in a barrel-like cavity, providing additional time for proper folding.

- Following their synthesis, all proteins are modified in various ways that alter their structure and function.
- The life span of intracellular proteins is largely determined by their susceptibility to proteolytic degradation by various pathways.
- The presence of certain internal sequences or Nterminal residues targets cytosolic proteins for addition of ubiquitin and subsequent proteolysis within a proteasome.

Genomics and Proteomics

Genomics is a discipline in genetics concerning the study of the genomes of organisms. The field includes intensive efforts to determine the entire DNA sequence of organisms and fine-scale genetic mapping efforts. The field also includes studies of intragenomic phenomena such as heterosis, epistasis, pleiotropy and other interactions between loci and alleles within the genome. In contrast, the investigation of the roles and functions of single genes is a primary focus of molecular biology or genetics and is a common topic of modern medical and biological research. Research of single genes does not fall into the definition of genomics unless the aim of this genetic, pathway, and functional information analysis is to elucidate its effect on, place in, and response to the entire genome's networks.

For the United States Environmental Protection Agency, "the term "genomics" encompasses a broader scope of scientific inquiry associated technologies than when genomics was initially considered. A genome is the sum total of all an individual organism's genes. Thus, genomics is the study of all the genes of a cell, or tissue, at the DNA (genotype), mRNA (transcriptome), or protein (proteome) levels."

History

The first genomes to be sequenced were those of a virus and a mitochondrion, and were done by Fred Sanger. His group established techniques of sequencing, genome mapping, data storage, and bioinformatic analyses in the 1970-1980s. A major branch of genomics is still concerned with sequencing the genomes of various organisms, but the knowledge of full genomes has created the possibility for the field of functional genomics, mainly concerned with patterns of gene expression during various conditions.

The most important tools here are microarrays and bioinformatics. Study of the full set of proteins in a cell type or tissue, and the changes during various conditions, is called proteomics. A related concept is materiomics, which is defined as the study of the material properties of biological materials (e.g. hierarchical protein structures and materials, mineralized biological tissues, etc.) and their effect on the macroscopic function and failure in their biological context, linking processes, structure and properties at multiple scales through a materials science approach. The actual term 'genomics' is thought to have been coined by Dr. Tom Roderick, a geneticist at the Jackson Laboratory (Bar Harbor, ME) over beer at a meeting held in Maryland on the mapping of the human genome in 1986.

In 1972, Walter Fiers and his team at the Laboratory of Molecular Biology of the University of Ghent (Ghent, Belgium) were the first to determine the sequence of a gene: the gene for Bacteriophage MS2 coat protein. In 1976, the team determined the complete nucleotide-sequence of bacteriophage MS2-RNA. The first DNA-based genome to be sequenced in its entirety was that of bacteriophage Φ-X174; (5,368 bp), sequenced by Frederick Sanger in 1977. The first free-living organism to be sequenced was that of *Haemophilus influenzae* (1.8 Mb) in 1995, and since then genomes are being sequenced at a rapid pace. As of September 2007, the complete sequence was known of about 1879 viruses, 577 bacterial species and roughly 23 eukaryote organisms, of which about half are fungi. Most of the bacteria whose genomes have been completely sequenced are problematic disease-causing agents, such as *Haemophilus influenzae.* Of the other sequenced species, most were chosen because they were well-studied model organisms or promised to become good models. Yeast (*Saccharomyces cerevisiae*) has long been an important model organism for the eukaryotic cell, while the fruit fly *Drosophila melanogaster* has been a very important tool (notably in early pre-molecular genetics). The worm *Caenorhabditis elegans* is an often used simple model for multicellular organisms. The zebrafish *Brachydanio rerio* is used for many developmental studies on the molecular level and the flower *Arabidopsis thaliana* is a model organism for flowering plants. The Japanese pufferfish (*Takifugu rubripes*) and the spotted green pufferfish (*Tetraodon nigroviridis*) are interesting because of their small and compact genomes, containing very little non-coding DNA compared to most species. The mammals dog (*Canis familiaris*), brown rat (*Rattus norvegicus*), mouse (*Mus musculus*), and chimpanzee (*Pan troglodytes*) are all important model animals in medical research.

Human Genomics

A rough draft of the human genome was completed by the Human Genome Project in early 2001, creating much fanfare. By 2007 the human sequence was declared "finished" (less than one error in 20,000 bases and all chromosomes assembled). Display of the results of the project required significant bioinformatics resources. The sequence of the human reference assembly can be explored using the UCSC Genome Browser.

Bacteriophage Genomics

Bacteriophages have played and continue to play a key role in bacterial genetics and molecular biology. Historically, they were used to define gene structure and gene regulation. Also the first genome to be sequenced was a bacteriophage. However, bacteriophage research did not lead the genomics revolution, which is clearly dominated by bacterial genomics.

Only very recently has the study of bacteriophage genomes become prominent, thereby enabling researchers to understand the mechanisms underlying phage evolution. Bacteriophage genome sequences can be obtained through direct sequencing of isolated bacteriophages, but can also be derived as part of microbial genomes. Analysis of bacterial genomes has shown that a substantial amount of microbial DNA consists of prophage sequences and prophage-like elements. A detailed database mining of these sequences offers insights into the role of prophages in shaping the bacterial genome.

Cyanobacteria Genomics

At present there are 24 cyanobacteria for which a total genome sequence is available. 15 of these cyanobacteria come from the marine environment.

These are six *Prochlorococcus* strains, seven marine *Synechococcus* strains, *Trichodesmium erythraeum* IMS101 and *Crocosphaera watsonii* WH8501. Several studies have demonstrated how tese sequences could be used very successfully to infer important ecological and physiological characteristics of marine cyanobacteria.

However, there are many more genome projects currently in progress, amongst those there are further *Prochlorococcus* and marine *Synechococcus* isolates, *Acaryochloris* and *Prochloron*, the N_2-fixing filamentous cyanobacteria *Nodularia spumigena*, *Lyngbya aestuarii* and *Lyngbya majuscula*, as well as bacteriophages infecting marine cyanobaceria.

Thus, the growing body of genome information can also be tapped in a more general way to address global problems by applying a comparative approach. Some new and exciting examples of progress in this field are the identification of genes for regulatory RNAs, insights into the evolutionary origin of photosynthesis, or estimation of the contribution of horizontal gene transfer to the genomes that have been analysed.

Proteomics

Proteomics is the large-scale study of proteins, particularly their structures and functions. Proteins are vital parts of living organisms, as they are the main components of the physiological metabolic pathways of cells. The term "proteomics" was first coined in 1997 to make an analogy with genomics, the study of the genes. The word "proteome" is a blend of "protein" and "genome", and was coined by Marc Wilkins in 1994 while working on the concept as a PhD student. The proteome is the entire complement of proteins, including the modifications made to a particular set of proteins, produced by an organism or system. This will vary with time and distinct requirements, or stresses, that a cell or organism undergoes.

Complexity of the Problem

After genomics, proteomics is considered the next step in the study of biological systems. It is much more complicated than genomics mostly because while an organism's genome is more or less constant, the proteome differs from cell to cell and from time to time. This is because distinct genes are expressed in distinct cell types. This means that even the basic set of proteins which are produced in a cell needs to be determined.

In the past this was done by mRNA analysis, but this was found not to correlate with protein content. It is now known that mRNA is not always translated into protein, and the amount of protein produced for a given amount of mRNA depends on the gene it is transcribed from and on the current physiological state of the cell. Proteomics confirms the presence of the protein and provides a direct measure of the quantity present.

Post-translational Modifications

Not only does the translation from mRNA cause differences, many proteins are also subjected to a wide variety of chemical modifications after translation. Many of these post-translational modifications are critical to the protein's function.

Phosphorylation

One such modification is phosphorylation, which happens to many enzymes and structural proteins in the process of cell signaling. The addition of a phosphate to particular amino acids—most commonly serine and threonine mediated by serine/threonine kinases, or more rarely tyrosine mediated by tyrosine kinases—causes a protein to become a target for binding or interacting with a distinct set of other proteins that recognize the phosphorylated domain.

Because protein phosphorylation is one of the most-studied protein modifications, many "proteomic" efforts are geared to determining the set of phosphorylated proteins in a particular cell or tissue-type under particular circumstances. This alerts the scientist to the signaling pathways that may be active in that instance.

Ubiquitination

Ubiquitin is a small protein that can be affixed to certain protein substrates by enzymes called E3 ubiquitin ligases. Determining which proteins are poly-ubiquitinated can be helpful in understanding how protein pathways are regulated. This is therefore an additional legitimate "proteomic" study. Similarly, once it is determined what substrates are ubiquitinated by each ligase, determining the set of ligases expressed in a particular cell type will be helpful.

Additional Modifications

Listing all the protein modifications that might be studied in a "Proteomics" project would require a discussion of most of biochemistry; therefore, a short list will serve here to illustrate the complexity of the problem. In addition to phosphorylation and ubiquitination, proteins can be subjected to (among others) methylation, acetylation, glycosylation, oxidation and nitrosylation. Some proteins undergo ALL of these modifications, often in time-dependent combinations, aptly illustrating the potential complexity one has to deal with when studying protein structure and function.

Distinct Proteins are Made under Distinct Settings

Even if one is studying a particular cell type, that cell may make different sets of proteins at different times, or under different conditions. Furthermore, as mentioned, any one protein can undergo a wide range of post-translational modifications. Therefore a "proteomics" study can become quite complex very quickly, even if the object of the study is very restricted. In more ambitious settings, such as when a biomarker for a tumour is sought - when the proteomics scientist is obliged to study sera samples from multiple cancer patients - the amount of complexity that must be dealt with is as great as in any modern biological project.

Limitations to Genomic Study

Scientists are very interested in proteomics because it gives a much better understanding of an organism than genomics. First, the level of transcription of a gene gives only a rough estimate of its level of expression into a protein. An mRNA produced in abundance may be degraded rapidly or translated inefficiently, resulting in a small amount of protein. Second, as mentioned above many proteins experience post-translational modifications that profoundly affect their activities; for example some proteins are not active until they become phosphorylated.

Methods such as phosphoproteomics and glycoproteomics are used to study post-translational modifications. Third, many transcripts give rise to more than one protein, through alternative splicing or alternative post-translational modifications. Fourth, many proteins form complexes with other proteins or RNA molecules, and only function in the presence of these other molecules. Finally, protein degradation rate plays an important role in protein content.

Methods of Studying Proteins

Determining Proteins which are Post-translationally Modified

One way in which a particular protein can be studied is to develop an antibody which is specific to that modification. For example, there are antibodies which only recognize certain proteins when they are tyrosine-phosphorylated, known as phospho-specific antibodies; also, there are antibodies specific to other modifications. These can be used to determine the set of proteins that have undergone the modification of interest.

For sugar modifications, such as glycosylation of proteins, certain lectins have been discovered which bind sugars. These too can be used.

A more common way to determine post-translational modification of interest is to subject a complex mixture of proteins to electrophoresis in "two-dimensions", which simply means that the proteins are electrophoresed first in one direction, and then in another... this allows small differences in a protein to be visualized by separating a modified protein from its unmodified form. This methodology is known as "two-dimensional gel electrophoresis". Recently, another approach has been developed called PROTOMAP which combines SDS-PAGE with shotgun proteomics to enable detection of changes in gel-migration such as those caused by proteolysis or post translational modification.

Determining the Existence of Proteins in Complex Mixtures

Classically, antibodies to particular proteins or to their modified forms have been used in biochemistry and cell biology studies. These are among the most common tools used by practicing biologists today. For more quantitative determinations of protein amounts, techniques such as ELISAs can be used.

For proteomic study, more recent techniques such as matrix-assisted laser desorption/ionization (MALDI) have been employed for rapid determination of proteins in particular mixtures and increasingly electrospray ionization (ESI).

Computational Methods in Studying Protein Biomarkers

Computational predictive models have shown that extensive and diverse feto-maternal protein trafficking occurs during pregnancy and can be readily detected non-invasively in maternal whole blood. This computational approach circumvented a major limitation, the abundance of maternal proteins interfering with the detection of fetal proteins, to fetal proteomic analysis of maternal blood.

Computational models can use fetal gene transcripts previously identified in maternal whole blood to create a comprehensive proteomic network of the term neonate. Such work shows that the fetal proteins detected in pregnant woman's blood originate from a diverse group of tissues and organs from the developing fetus.

The proteomic networks contain many biomarkers that are proxies for development and illustrate the potential clinical application of this technology as a way to monitor normal and abnormal fetal development. An information theoretic framework has also been introduced for biomarker discovery, integrating biofluid and tissue information.

This new approach takes advantage of functional synergy between certain biofluids and tissues with the potential for clinically significant findings not possible if tissues and biofluids were considered individually.

By conceptualizing tissue-biofluid as information channels, significant biofluid proxies can be identified and then used for guided development of clinical diagnostics. Candidate biomarkers are then predicted based on information transfer criteria across the tissue-biofluid channels. Significant biofluid-tissue relationships can be used to prioritize clinical validation of biomarkers.

Establishing Protein-protein Interactions

Most proteins function in collaboration with other proteins, and one goal of proteomics is to identfy which proteins interact. This is especially useful in determining potential partners in cell signaling cascades. Several methods are available to probe protein-protein interactions. The traditional method is yeast two-hybrid analysis. New methods include protein microarrays, immunoaffinity chromatography followed by mass spectrometry, dual polarisation interferometry, Microscale Thermophoresis and experimental methods such as phage display and computational methods

Practical Applications of Proteomics

One of the most promising developments to come from the study of human genes and proteins has been the identification of potential new drugs for the treatment of disease. This relies on geome and proteome information to identify proteins associated with a disease, which computer software can then use as targets for new drugs. For example, if a certain protein is implicated in a disese, its 3D structure provides the information to design drugs to interfere with the action of the protein. A molecule that fits the active site of an enzyme, but cannot be released by the enzyme, will inactivate the enzyme. This is the basis of new drug-discovery tools, which aim to find new drugs to inactivate proteins involved in disease. As genetic differences among individuals are found, researchers expect to use these techniques to develop personalized drugs that are more effectiv

Chapter 9

Organisation of Genome

Genome organization refers to the sequential, not the structural organization of the genome. Besides the coding exons, the non-coding DNA in Eukaryotes may fall in the following classes:

- Introns. They are DNA sequences inserted between the exons and found in the ORF.
- Pseudogenes. 'Dead', non-functional copies of genes present elsewhere in the genome, but no longer of any use.
- Retropseudogenes. Like pseudogenes, but have been processed, i.e. lack introns. Produced by the action of reverse transcriptase (RT) on mRNA, and subsequent incorporation of the cDNA into the genome.
- Transposons. Jumping genes, which splice themselves in and out of the genome (in DNA form) randomly, by the action of transposase.
- Retrotransposons. Transcribed into an mRNA, which encodes an RT enzyme, which then copies the mRNA back to DNA and incorporates it into the genome.

Infact in humans only 1.5% of the entire genome length corresponds to coding DNA. This 1.5% codes for about 27,000 genes which in turn code for proteins that are responsible for all the cellular processes.

Genome Analysis?

Genome analysis entails the prediction of genes in uncharacterized genomic sequences. The 21st century has seen the announcement of the draft version of the human genome sequence. Model organisms have been sequenced in both the plant and animal kingdoms. However, the pace of genome annotation is not matching the pace of genome sequencing. Experimental genome annotation is slow and time consuming. The demand is to be able to develop computational tools for gene prediction.

Computational Gene prediction is relatively simple for the prokaryotes where all the genes are converted into the corresponding mRNA and then into proteins. The process is more complex for eukaryotic cells where the coding DNA sequence is interrupted by random sequences called introns.

Some of the questions which biologists want to answer today are:

- Given a DNA sequence, what part of it codes for a protein and what part of it is junk DNA.
- Classify the junk DNA as intron, untranslated region, transposons, dead genes, regulatory elements etc.
- Divide a newly sequenced genome into the genes (coding) and the non-coding regions.

The importance of genome analysis can be understood by comparing the human and chimpanzee genomes. The chimp and human genomes vary by an average of just 2% i.e. just about 160 enzymes.

A complete genome analysis of the two genomes would give a strong insight into the various mechanisms responsible for the differences.

Arabidopsis and Humans have the same number of genes, though the Arabidopsis genome is around 250 times smaller than humans. How is that ?

The human genome has a lot of junk DNA, specifically transposons and mobile genetic elements. This increases the size of the human genome, though the number of genes is only 27,000.

However, the number of protein products in humans is significantly higher. Many of the sequenced human genes have alternative splice products. In addition, several other processes (e.g. signal transduction) proceed via further protein modifications, such as Glycosylation. Therefore, the number of human protein products could far exceed the number of genes. Why do plants have such bulky genomes when they are not as complex as some of the higher eukaryotes ? This is mostly due to two factors: the ability of plants to duplicate their genomes in order to reproduce (a process known as polyploidization) and the susceptibility of plants to mobile genetic elements.

- Polyploidization allows plants to more easily form hybrids when pollen and ova from different species come together. The result of such hybridization events are plants with genomes that are the sum of the two parent genome sizes (as opposed to half of one parent's genome and half of the other parent's genome as in normal sexual reproduction.

- Also, in case of plants, it is fairly common to observe insertion of transposable elements in intergenic regions. This also explains the difference in the sizes of plant genomes among themselves as well.

What is Open Reading Frame (ORF) ?

The region of the nucleotide sequences from the start codon (ATG) to the stop codon is called the Open Reading frame. Gene finding in organism specially prokaryotes starts form searching for an open reading frames (ORF). An ORF is a sequence of DNA that starts with start codon "ATG" (not always) and ends with any of the three termination codons (TAA, TAG, TGA).

Depending on the starting point, there are six possible ways (three on forward strand and three on complementary strand) of translating any nucleotide sequence into amino acid sequence according to the genetic code.These are called reading frames. While eukaryotic gene finding is altogether a different task as the eukaryotic genes are not continuous and interrupted by intervening noncoding sequences called 'introns'. Moreover organization of genetic information in eukaryotes and prokaryotes is different.

What is Coding Sequence(CDS)? How is it Different from the ORF?

The Coding Sequence (CDS) is the actual region of DNA that is translated to form proteins. While the ORF may contain introns as well, the CDS refers to those nucleotides(concatenated exons) that can be divided into codons which are actually translated into amino acids by the ribosomal translation machinery. In Prokaryotes the ORF and the CDS are the same.

DNA Replication

Before a cell divides, its DNA is replicated (duplicated.) Because the two strands of a DNA molecule have complementary base pairs, the nucleotide sequence of each strand automatically supplies the information needed to produce its partner.

If the two strands of a DNA molecule are separated, each can be used as a pattern or template to produce a complementary strand.

Each template and its new complement together then form a new DNA double helix, identical to the original.

Before replication can occur, the length of the DNA double helix about to be copied must be unwound.

In addition, the two strands must be separated, much like the two sides of a zipper, by breaking the weak hydrogen bonds that link the paired bases. Once the DNA strands have been unwound, they must be held apart to expose the bases so that new nucleotide partners can hydrogen-bond to them.

The enzyme DNA polymerase then moves along the exposed DNA strand, joining newly arrived nucleotides into a new DNA strand that is complementary to the template.

Each cell contains a family of more than thirty enzymes to insure the accurate replication of DNA.

Primers

Though DNA polymerase can elongate a polynucleotide strand by adding new nucleotides, it cannot start a strand from scratch because it can only bond new nucleotides to a free sugar (3') end of a nucleotide chain. DNA polymerase requires the assistance of a primer, a previously existing short strand of DNA (or RNA) that is complementary to the first part of the DNA segment being copied.

This small strand of nucleotides anneals (binds) by complementary base pairing to the beginning of the area being copied. With the primer in place, DNA polymerase is then able to continue adding the rest of the pairs of the segment until a new double strand of DNA is completed. Primers are formed from free nucleotides in the cell by enzymes called DNA primases.

Replication Occurs Differently on Antiparallel Strands of DNA

That nucleotides can be added only to the sugar or 3' end of the growing complementary chain presents no problem for the side of the DNA chain opening at its phosphate or 5' end. The primer that binds to the first few exposed bases will end with a sugar (3') where the phosphate of a new nucleotide can be attached. From there on, DNA polymerase can continuously synthesize the growing complementary strand. This strand of DNA is called the leading strand. A different challenge faces DNA polymerase when the complementary side of the DNA molecule begins unzipping from its sugar (3') toward its phosphate (5') end. A primer of complementary molecules attaching to the opening end of this chain would have a phosphate not a sugar at its exposed end so that new nucleotides could not be joined. To get around this problem, this strand is synthesized in small pieces backward from the overall direction of replication.

This strand is called the lagging strand. The short segments of newly assembled DNA from which the lagging strand is built are called Okazaki fragments. As replication proceeds and nucleotides are added to the 3' end of the Okazaki fragments, they come to meet each other. The primer fragments are then booted out by enzymes and replaced by appropriate DNA nucleotides. The whole thing is then stitched together by another enzyme called DNA ligase.

DNA Damage, Repair, and Recombination

DNA repair refers to a collection of processes by which a cell identifies and corrects damage to the DNA molecules that encode its genome. In human cells, both normal metabolic activities and environmental factors such as UV light and radiation can cause DNA damage, resulting in as many as 1 million individual molecular lesions per cell per day.

Many of these lesions cause structural damage to the DNA molecule and can alter or eliminate the cell's ability to transcribe the gene that the affected DNA encodes. Other lesions induce potentially harmful mutations in the cell's genome, which affect the survival of its daughter cells after it undergoes mitosis. As a consequence, the DNA repair process is constantly active as it responds to damage in the DNA structure. When normal repair processes fail, and when cellular apoptosis does not occur, irreparable DNA damage may occur, including double-strand breaks and DNA crosslinkages.

The rate of DNA repair is dependent on many factors, including the cell type, the age of the cell, and the extracellular environment. A cell that has accumulated a large amount of DNA damage, or one that no longer effectively repairs damage incurred to its DNA, can enter one of three possible states:

1. an irreversible state of dormancy, known as senescence
2. cell suicide, also known as apoptosis or programmed cell death
3. unregulated cell division, which can lead to the formation of a tumor that is cancerous.

The DNA repair ability of a cell is vital to the integrity of its genome and thus to its normal functioning and that of the organism. Many genes that were initially shown to influence life span have turned out to be involved in DNA damage repair and protection. Failure to correct mlecular lesions in cells that form gametes can introduce mutations into the genomes of the offspring and thus influence the rate of evolution.

DNA Damage

DNA damage, due to environmental factors and normal metabolic processes inside the cell, occurs at a rate of 1,000 to 1,000,000 molecular lesions per cell per day. While this constitutes only 0.000165% of the human genome's approximately 6 billion bases (3 billion base pairs), unrepaired lesions in critical genes (such as tumor suppressor genes) can impede a cell's ability to carry out its function and appreciably increase the likelihood of tumour formation.

The vast majority of DNA damage affects the primary structure of the double helix; that is, the bases themselves are chemically modified. These modifications can in turn disrupt the molecules' regular helical structure by introducing non-native chemical bonds or bulky adducts that do not fit in the standard double helix.

Unlike proteins and RNA, DNA usually lacks tertiary structure and therefore damage or disturbance does not occur at that level. DNA is, however, supercoiled and wound around "packaging" proteins called histones (in eukaryotes), and both superstructures are vulnerable to the effects of DNA damage.

Sources of Damage

DNA damage can be subdivided into two main types:

1. endogenous damage such as attack by reactive oxygen species produced from normal metabolic byproducts (spontaneous mutation), especially the process of oxidative deamination
    - also includes replication errors.
2. exogenous damage caused by external agents such as
    - ultraviolet [UV 200-300nm] radiation from the sun
    - other radiation frequencies, including x-rays and gamma rays
    - hydrolysis or thermal disruption
    - certain plant toxins
    - human-made mutagenic chemicals, especially aromatic compounds that act as DNA intercalating agents
    - cancer chemotherapy and radiotherapy
    - viruses.

The replication of damaged DNA before cell division can lead to the incorporation of wrong bases opposite damaged ones.

Daughter cells that inherit these wrong bases carry mutations from which the original DNA sequence is unrecoverable (except in the rare case of a back mutation, for example, through gene conversion).

Types of Damage

There are five main types of damage to DNA due to endogenous cellular processes:

1. *oxidation* of bases [e.g. 8-oxo-7,8-dihydroguanine (8-oxoG)] and generation of DNA strand interruptions from reactive oxygen species,
2. *alkylation* of bases (usually methylation), such as formation of 7-methylguanine, 1-methyladenine, 6-O-Methylguanine
3. *hydrolysis* of bases, such as deamination, depurination, and depyrimidination.
4. "bulky adduct formation" (i.e., benzo[a]pyrene diol epoxide-dG adduct)
5. *mismatch* of bases, due to errors in DNA replication, in which the wrong DNA base is stitched into place in a newly forming DNA strand, or a DNA base is skipped over or mistakenly inserted.

Damage caused by exogenous agents comes in many forms. Some examples are:

1. *UV-B light* causes crosslinking between adjacent cytosine and thymine bases creating *pyrimidine dimers*. This is called direct DNA damage.
2. *UV-A light* creates mostly free radicals. The damage caused by free radicals is called indirect DNA damage.
3. *Ionizing radiation* such as that created by radioactive decay or in *cosmic rays* causes breaks in DNA strands. Low-level ionizing radiation may induce irreparable DNA damage (leading to replicational and transcriptional errors needed for neoplasia or may trigger viral interactions) leading to pre-mature aging and cancer.
4. *Thermal disruption* at elevated temperature increases the rate of depurination (loss of purine bases from the DNA backbone) and single-strand breaks. For example, hydrolytic depurination is seen in the thermophilic bacteria, which grow in hot springs

at 40-80 °C. The rate of depurination (300 purine residues per genome per generation) is too high in these species to be repaired by normal repair machinery, hence a possibility of an adaptive response cannot be ruled out.

5. *Industrial chemicals* such as vinyl chloride and hydrogen peroxide, and environmental chemicals such as polycyclic hydrocarbons found in smoke, soot and tar create a huge diversity of DNA adducts- ethenobases, oxidized bases, alkylated phosphotriesters and Crosslinking of DNA just to name a few.

UV damage, alkylation/methylation, X-ray damage and oxidative damage are examples of induced damage. Spontaneous damage can include the loss of a base, deamination, sugar ring puckering and tautomeric shift.

Nuclear versus Mitochondrial DNA Damage

In human cells, and eukaryotic cells in general, DNA is found in two cellular locations - inside the nucleus and inside the mitochondria. Nuclear DNA (nDNA) exists as chromatin during non-replicative stages of the cell cycle and is condensed into aggregate structures known as chromosomes during cell division. In either state the DNA is highly compacted and wound up around bead-like proteins called histones.

Whenever a cell needs to express the genetic information encoded in its nDNA the required chromosomal region is unravelled, genes located therein are expressed, and then the region is condensed back to its resting conformation. Mitochondrial DNA (mtDNA) is located inside mitochondria organelles, exists in multiple copies, and is also tightly associated with a number of proteins to form a complex known as the nucleoid.

Inside mitochondria, reactive oxygen species (ROS), or free radicals, byproducts of the constant production of adenosine triphosphate (ATP) via oxidative phosphorylation, create a highly oxidative environment that is known to damage mtDNA. A critical enzyme in counteracting the toxicity of these species is superoxide dismutase, which is present in both the mitochondria and cytoplasm of eukaryotic cells.

Senescence and Apoptosis

Senescence, an irreversible state in which the cell no longer divides, is a protective response to the shortening of the chromosome ends. The telomeres are long regions of repetitive noncoding DNA that cap chromosomes and undergo partial degradation each time a cell undergoes division. In contrast, quiescence is a reversible state of cellular dormancy that is unrelated to genome damage. Senescence in cells may serve as a functional alternative to apoptosis in cases where the physical presence of a cell for spatial reasons is required by the organism, which serves as a "last resort" mechanism to prevent a cell with damaged DNA from replicating inappropriately in the absence of pro-growth cellular signaling.

Unregulated cell division can lead to the formation of a tumour, which is potentially lethal to an organism. Therefore, the induction of senescence and apoptosis is considered to be part of a strategy of protection against cancer.

DNA Damage and Mutation

It is important to distinguish between DNA damage and mutation, the two major types of error in DNA. DNA damages and mutation are fundamentally different. Damages are physical abnormalities in the DNA, such as single- and double-strand breaks, 8 hydroxydeoxyguanosine residues, and polycyclic aromatic hydrocarbon adducts.

DNA damages can be recognized by enzymes, and, thus, they can be correctly repaired if redundant information, such as the undamaged sequence in the complementary DNA strand or in a homologous chromosome, is available for copying. If a cell retains DNA damage, transcription of a gene can be prevented, and, thus, translation into a protein will also be blocked. Replication may also be blocked and/or the cell may die.

In contrast to DNA damage, a mutation is a change in the base sequence of the DNA. A mutation cannot be recognized by enzymes once the base change is present in both DNA strands, and, thus, a mutation cannot be repaired. At the cellular level, mutations can cause alterations in protein function and regulation. Mutations are replicated when the cell replicates.

In a population of cells, mutant cells will increase or decrease in frequency according to the effects of the mutation on the ability of the cell to survive and reproduce. Although distinctly different from each other, DNA damages and mutations are related because DNA damages often cause errors of DNA synthesis during replication or repair; these errors are a major source of mutation.

Given these properties of DNA damage and mutation, it can be seen that DNA damages are a special problem in non-dividing or slowly dividing cells, where unrepaired damages will tend to accumulate over time. On the other hand, in rapidly dividing cells, unrepaired DNA damages that do not kill the cell by blocking replication will tend to cause replication errors and thus mutation. The great majority of mutations that are not neutral in their effect are deleterious to a cell's survival. Thus, in a population of cells comprising a tissue with replicating cells, mutant cells will tend to be lost.

However, infrequent mutations that provide a survival advantage will tend to clonally expand at the expense of neighboring cells in the tissue. This advantage to the cell is disadvantageous to the whole organism, because such mutant cells can give rise to cancer. Thus, DNA damages in frequently dividng cells, because they give rise to mutations, are a prominent cause of cancer. In contrast, DNA damages in infrequently dividing cells are likely a prominent ause of aging.

DNA Repair Mechanisms

Cells cannot function if DNA damage corrupts the integrity and accessibility of essential information in the genome (but cells remain superficially functional when so-called "non-essential" genes are missing or damaged). Depending on the type of damage inflicted on the DNA's double helical structure, a variety of repair strategies have evolved to restore lost information.

If possible, cells use the unmodified complementary strand of the DNA or the sisterchromatid as a template to recover the original information. Without access to a template, cells use an error-prone recovery mechanism known as translesion synthesis as a last resort. Damage to DNA alters the spatial configuration of the helix, and such alteratins can be detected by the cell. Once damage is localized, specific DNA repair molecules bind at or near the site of damage, inducing other molecules to bind and form a complex that enables he actual repair to take place.

Direct Reversal

Cells are known to eliminate threetypes of damage to their DNA by chemically reversing it. These mechanisms do not require a template, since the types of damage they counteract can occur in only one of the four bases. Such direct reversal mechanisms are specific to the type of damage incurred and do not involve breakage of the phosphodiester backbone.

The formation of pyrimidine dimers upon irradiatio with UV light results in an abnormal covalent bond between adjacent pyrimidine bases. The photoreactivation process directly reverses this damage by the action of the enzyme photolyase, whose activation is obligately dependent on energy absorbed frm blue/UV light (300–500 nm wavelength) to promote catalysis. Another type of damage, methylation of guanine bases, is directly reversed by the protein methyl guanine methl transferase (MGMT), the bacterial equivalent of which is called ogt. This is an expensive process because each MGMT molecule can be used only once; that is, the reaction is stoichiometric rather than catalytic.

A generalized response to methylating agents in bacteria is known as the adaptive response and confers a level of resistance to alkylating agents upon sustained exposure by upregulation of alkylation repair enzymes. The third type of DNA damage reversed by cells is certain methylation of the bases cytosine and adenine.

Single-strand Damage

When only one of the two strands of a double helix has a defect, the other strand can be used as a template to guide the correction of the damaged strand. In order to repair damage to one of the two paired molecules of DNA, there exist a number of excision repair mechanisms that remove the damaged nuclotide and replace it with an undamaged nucleotide complementary to that found in the undamagd DNA strand.

1. Base excision rpair (BER), which repairs damage to a single base caused by oxidation, alkylation, hydrolysis, or deamination. The damaged base is removed by a DNA glycosylase. The "missing tooth" is then recognised by an enzyme called AP endonuclease, which cuts the Phosphodiester bond. The missing part is then resynthesized by a DNA polymerase, and a DNA ligase performs the final nick-sealing step.
2. Nucleotide excision repair (NER), which recognizes bulky, helix-distorting lesions such as pyrimidine dimers and 6,4 photoproducts. A specialized form of NER known as tanscription-coupled repair deploys NER enzymes to genes that are being actively transcribed.
3. Mismatch repair (MMR), which corrects errors of DNAreplication and recombination that result in mispaired (but undamaged) nucleotides.

Double-strand Breaks

Double-srand breaks, in which both strands in the double helix are severed, are particularly hazardous to the cell bcause they can lead to genome rearrangements. Three mechanisms exist to repair double-strand breaks (DSBs): non-homologous end joining (NHEJ), microhomology-mediated end joining (MMEJ), and homologous recombination. PVN Acharya noted that double-strand breaks and a "cross-linkage joining both strands at the same point is irreparable because neither strand can then serve as a template for repair. The cell will die in the nxt mitosis or in some rare instances, mutate."

In NHEJ, DNA Ligase IV, aspecialized DNA ligase tha forms a complex with the cofactor XRCC4, directly joins the two ends. To guide accurte repair, NHEJ relies on short homologous sequences called microhomologies present on the single-stranded tails of the DNA ends to be joined. If these overhangs are compatible, repairis usually accurate. NHEJ can also introduce mutations during repair.

Loss of damaged nucleotides at the break site can lead to deletions, and joining of nonmatching termini forms translocations. NHEJ is especially important before the cell has replicated its DNA, since there is no template available for repair by homologous recombination. There are "backup" NHEJ pathways in higher eukaryotes. Besides its role as a genome caretaker, NHEJ is required for joining hairpin-capped double-strand breaks induced during V(D)J recombination,the process that generates diversity in B-cell and T-cell receptors in the vertebrate immune system.

Homologous recombination requires the presence of an identical or nearly idntical sequence to be used as a template for repair of the break. The enzymatic machinery responsible for this repair process is nearly identical to the machinery responsible for chromosomal crossover during meiosis. This pathway allows a damaged chromosome to be repaired using a sister chromatid (available in G2 after DNA replication) or a homologous chromosome as a template. DSBs caused by the replication machinery attempting to synthesize across a single-strand break or unrepaired lesion cause collapse of the replication fork and are typically repaired by recombination.

Topoisomerases introduce both single- and double-strand breaks in the course of changing the DNA's state of supercoiling, which is especially common in regions near an open replication fork. Such breaks are not considered DNA damage because they are a natural intermediate in the topoisomerase biochemical mechanism and are immediately repaired by the enzymes that created them.

A team of French researchers bombarded *Deinococcus radiodurans* to study the mechanism of double-strand break DNA repair in that organism. At least two copies of the genome, with random DNA breaks, can form DNA fragments through annealng. Partially overlapping fragments are then use for synthesis of homologous regions through a moving D-loop that can continue extension until they find complementary partner strands. In the final step there is crossover by means of RecA-dependent homologous recombination.

Translesion Synthesis

Translesion synthesis is a DNA damage tolerance process that allows the DNA replication machinery to replicate past DNA lesions such as thymine dimers or AP sites. It involves switching out regular DNA polymerases for specialized translesion polymerases (i.e. DNA polymerase IV or V, from the Y Polymerase family), often with larger active sites that can facilitate the insertion of bases opposite damaged nucleotides.

The polymerase switching is thought to be mediated by, among other factors, the post-translational modification of the replication processivity factor PCNA. Translesion synthesis polymerases often have low fidelity (high propensity to insert wrong bases) on undamaged templates relative to regular polymerases. However, many are extremely efficient at inserting correct bases opposite specific types of damage. For example, Pol ç mediates error-free bypass of lesions induced by UV irradiation, whereas Pol æ introduces mutations at these sites. From a cellular perspective, risking the introduction of point mutations during translesion synthesis may be preferable to resorting to more drastic mechanisms of DNA repair, which may cause gross chromosomal aberrations or cell death.

In short, the process involves specialized polymerases either bypassing or repairing lesions at locations of stalled DNA replication. A bypass platform is provided to these polymerases by Proliferating cell nuclear antigen (PCNA). Under normal circumstances, PCNA bound to polymerases replicates the DNA. At a site of lesion, PCNA is ubiquitinated, or modified, by the RAD6/RAD18 proteins to provide a platform for the specialized polymerases to bypass the lesion and resume DNA replication.

Global Response to DNA Damage

Cells exposed to ionizing radiation, ultraviolet light or chemicals are prone to acquire multiple sites of bulky DNA lesions and double-strand breaks. Moreover, DNA damaging agents can damage other biomolecules such as proteins, carbohydrates, lipids, and RNA. The accumulation of damage, to be specific, double-strand breaks or adducts stalling the replication forks, are among known stimulation signals for a global response to DNA damage. The global response to damage is an act directed toward the cells' own preservation and triggers multiple pathways of macromolecular repair, lesion bypass, tolerance, or apoptosis. The common features of global response are induction of multiple genes, cell cycle arrest, and inhibition of cell division.

DNA Damage Checkpoints

After DNA damage, cell cycle checkpoints are activated. Checkpoint activation pauses the cell cycle and gives the cell time to repair the damage before continuing to divide. DNA damage checkpoints occur at the G1/ and G2/M boundaries. An intra-S checkpoint also exists. Checkpoint activation is controlled by two master kinases, ATM and ATR. ATM responds to DNA double-strand breaks and disruptions in chromatin structure, whereas ATR primarily responds to stalled replication forks. These kinases phosphorylate downstream targets in a signal transduction cascade, eventually leading to cell cycle arrest. A class of checkpoint mediator proteins including BRCA1, MDC1, and 53BP1 has also been identified. These proteins seem to be required for transmitting the checkpoint activation signal to downstream proteins.

P53 is an important downstream target of ATM and ATR, as it is required for inducing apoptosis following DNA damage. At the G1/S checkpoint, p53 functions by deactivating the CDK2/cyclin E complex. Similarly, p21 mediates the G2/M checkpoint by deactivating the CDK1/cyclin B complex.

The Prokaryotic SOS Response

The SOS response is the term used to describe changes in gene expression in *Escherichia coli* and other bacteria in response to extensive DNA damage. The prokaryotic SOS system is regulated by two key proteins: LexA and RecA. The LexA homodimer is a transcriptional repressor that binds to operator sequences commonly referred to as SOS boxes.

In *Escherichia coli* it is known that LexA regulates transcription of approximately 48 genes including the lexA and recA genes. The SOS response is known to be widespread in the Bacteria domain, but it is mostly absent in some bacterial phyla, like the Spirochetes.

The most common cellular signals activating the SOS response are regions of single-stranded DNA (ssDNA), arising from stalled replication forks or double-strand breaks, which are processed by DNA helicase to separate the two DNA strands. In the initiation step, RecA protein binds to ssDNA in an ATP hydrolysis driven reaction creating

RecA–ssDNA filaments. RecA–ssDNA filaments activate LexA autoprotease activity, which ultimately leads to cleavage of LexA dimer and subsequent LexA degradation. The loss of LexA repressor induces transcription of the SOS genes and allows for further signal induction, inhibition of cell division and an increase in levels of proteins responsible for damage processing.

In *Escherichia coli*, SOS boxes are 20-nucleotide long sequences near promoters with palindromic structure and a high degree of sequence conservation. In other classes and phyla, the sequence of SOS boxes varies considerably, with different length and cmposition, but it is always highly conserved and one of the strongest short signals in the genome. The high information content of SOS boxes permits differential binding of LexA to different promoters and allows for timing of the SOS response.

The lesion repair genes are induced at the beginning of SOS response. The error-prone translesion polymerases, for example, UmuCD'2 (also called DNA polymerase V), are induced later on as a last resort. Once the DNA damage is repaired or bypassed using polymerases or through recombination, the amount of single-stranded DNA in cells is decreased, lowering the amounts of RecA filaments decreases cleavage activity of LexA homodimer, which then binds to the SOS boxes near promoters and restores normal gene expression.

Eukaryotic Transcriptional Responses to DNA Damage

Eukaryotic cells exposed to DNA damaging agents also activate important defensive pathways by inducing multiple proteins involved in DNA repair, cell cycle checkpoint control, protein trafficking and degradation.

Such genome wide transcriptional response is very complex and tightly regulated, thus allowing coordinated global response to damage. Exposure of yeast *Saccharomyces cerevisiae* to DNA damaging agents results in overlapping but distinct transcriptional profiles. Similarities to environmental shock response indicates that a general global stress response pathway exist at the level of transcriptional activation.

In contrast, different human cell types respond to damage differently indicating an absence of a common global response. The probable explanation for this difference between yeast and human cells may be in the heterogeneity of mammalian cells. In an animal different types of cells are distributed among different organs that have evolved different sensitivities to DNA damage. In general global response to DNA damage involves expression of multiple genes responsible for postreplication repair, homologous recombination, nucleotide excision repair, DNA damage checkpoint, global transcriptional activation, genes controlling mRNA decay, and many others. A large amount of damage to a cell leaves it with an important decision: undergo apoptosis and die, or survive at the cost of living with a modified genome. An increase in tolerance to damage can lead to an increased rate of survival that will allow a greater accumulation of mutations. Yeast Rev1 and human polymerase ç are members of [Y family translesion DNA polymerases present during global response to DNA damage and are responsible for enhanced mutagenesis during a global response to DNA damage in eukaryotes.

DNA Repair and Aging

Pathological Effects of Poor DNA Repair

DNA repair rate is an important determinant of cell pathology.

Experimental animals with genetic deficiencies in DNA repair often show decreased life span and increased cancer incidence. For example, mice deficient in the dominant NHEJ pathway and in telomere maintenance mechanisms get lymphoma and infections more often, and, as a consequence, have shorter lifespans than wild-type mice.

In similar manner, mice deficient in a key repair and transcription protein that unwinds DNA helices have premature onset of aging-related diseases and consequent shortening of lifespan. However, not every DNA repair deficiency creates exactly the predicted effects; mice deficient in the NER pathway exhibited shortened life span without correspondingly higher rates of mutation.

If the rate of DNA damage exceeds the capacity of the cell to repair it, the accumulation of errors can overwhelm the cell and result in early senescence, apoptosis, or cancer. Inherited diseases associated with faulty DNA repair functioning result in premature aging, increased sensitivity to carcinogens, and correspondingly increased cancer risk.

On the other hand, organisms with enhanced DNA repair systems, such as *Deinococcus radiodurans*, the most radiation-resistant known organism, exhibit remarkable resistance to the double-strand break-inducing effects of radioactivity, likely due to enhanced efficiency of DNA repair and especially NHEJ.

Longevity and Caloric Restriction

A number of individual genes have been identified as influencing variations in life span within a population of organisms. The effects of these genes is strongly dependent on the environment, in particular, on the organism's diet. Caloric restriction reproducibly results in extended lifespan in a variety of organisms, likely via nutrient sensing pathways and decreased metabolic rate.

The molecular mechanisms by which such restriction results in lengthened lifespan are as yet unclear; however, the behavior of many genes known to be involved in DNA repair is altered under conditions of caloric restriction. For example, increasing the gene dosage of the gene SIR-2, which regulates DNA packaging in the nematode worm *Caenorhabditis elegans*, can significantly extend lifespan. The mammalian homolog of SIR-2 is known to induce downstream DNA repair factors involved in NHEJ, an activity that is especially promoted under conditions of caloric restriction. Caloric restriction has been closely linked to the rate of base excision repair in the nuclear DNA of rodents, although similar effects have not been observed in mitochondrial DNA.

It is interesting to note that the *C. elegans* gene AGE-1, an upstream effector of DNA repair pathways, confers dramatically extended life span under free-feeding conditions but leads to a decrease in reproductive fitness under conditions of caloric restriction. This observation supports the pleiotropy theory of the biological origins of aging, which suggests that genes conferring a large survival advantage early in life will be selected for even if they carry a corresponding disadvantage late in life.

Medicine and DNA Repair Modulation

Hereditary DNA Repair Disorders: Defects in the NER mechanism are responsible for several genetic disorders, including:

- Xeroderma pigmentosum: hypersensitivity to sunlight/UV, resulting in increased skin cancer incidence and premature aging
- Cockayne syndrome: hypersensitivity to UV and chemical agents
- Trichothiodystrophy: sensitive skin, brittle hair and nails.

Mental retardation often accompanies the latter two disorders, suggesting increased vulnerability of developmental neurons.

Other DNA repair disorders include:

- Werner's syndrome: premature aging and retarded growth
- Bloom's syndrome: sunlight hypersensitivity, high incidence of malignancies (especially leukemias).

- Ataxia telangiectasia: sensitivity to ionizing radiation and some chemical agents.

All of the above diseases are often called "segmental progerias" ("accelerated aging diseases") because their victims appear elderly and suffer from aging-related diseases at an abnormally young age, while not manifesting all the symptoms of old age.

Other diseases associated with reduced DNA repair function include Fanconi's anemia, hereditary breast cancer and hereditary colon cancer.

DNA Repair and Cancer

Inherited mutations that affect DNA repair genes are strongly associated with high cancer risks in humans. Hereditary nonpolyposis colorectal cancer (HNPCC) is strongly associated with specific mutations in the DNA mismatch repair pathway. BRCA1 and BRCA2, two famous mutations conferring a hugely increased risk of breast cancer on carriers, are both associated with a large number of DNA repair pathways, especially NHEJ and homologous recombination.

Cancer therapy procedures such as chemotherapy and radiotherapy work by overwhelming the capacity of the cell to repair DNA damage, resulting in cell death. Cells that are most rapidly dividing - most typically cancer cells - are preferentially affected. The side-effect is that other non-cancerous but rapidly dividing cells such as stem cells in the bone marrow are also affected. Modern cancer treatments attempt to localize the DNA damage to cells and tissues only associated with cancer, either by physical means (concentrating the therapeutic agent in the region of the tumour) or by biochemical means (exploiting a feature unique to cancer cells in the body).

DNA Repair and Evolution

The basic processes of DNA repair are highly conserved among both prokaryotes and eukaryotes and even among bacteriophage (viruses that infect bacteria); however, more complex organisms with more complex genomes have correspondingly more complex repair mechanisms. The ability of a large number of protein structural motifs to catalyze relevant chemical reactions has played a significant role in the elaboration of repair mechanisms during evolution. For an extremely detailed review of hypotheses relating to the evolution of DNA repair.

The fossil record indicates that single-cell life began to proliferate on the planet at some point during the Precambrian period, although exactly when recognizably modern life first emerged is unclear. Nucleic acids became the sole and universal means of encoding genetic information, requiring DNA repair mechanisms that in their basic form have been inherited by all extant life forms from their common ancestor.

The emergence of Earth's oxygen-rich atmosphere (known as the "oxygen catastrophe") due to photosynthetic organisms, as well as the presence of potentially damaging free radicals in the cell due to oxidative phosphorylation, necessitated the evolution of DNA repair mechanisms that act specifically to counter the types of damage induced by oxidative stress.

Rate of Evolutionary Change

On some occasions, DNA damage is not repaired, or is repaired by an error-prone mechanism that results in a change from the original sequence. When this occurs, mutations may propagate into the genomes of the cell's progeny. Should such an event occur in a germ line cell that will eventually produce a gamete, the mutation has the potential to be passed on to the organism's offspring.

The rate of evolution in a particular species (or, in a particular gene) is a function of the rate of mutation. As a consequence, the rate and accuracy of DNA repair mechanisms have an influence over the process of evolutionary change.

Senescence

Senescence or biological aging is the change in the biology of an organism as it ages after its maturity. Such changes range from those affecting its cells and their function to that of the whole organism. There are a number of theories as to why senescence occurs, including ones that claim it is programmed by gene expression changes and that it is the accumulative damage of biological processes.

The word *senescence* is derived from the Latin word *senex*, meaning *old man*, *old age*, or *advanced in age*.

Cellular Senescence

Cellular senescence is the phenomenon by which normal diploid cells lose the ability to divide, normally after about 50 cell divisions in vitro. Some cells become senescent after fewer replications cycles as a result of DNA double strand breaks, toxins, etc. This phenomenon is also known as "replicative senescence", the "Hayflick phenomenon", or the Hayflick limit in honour of Dr. Leonard Hayflick who was the first to publish this information in 1965. In response to DNA damage (including shortened telomeres), cells either age or self-destruct (apoptosis, programmed cell death) if the damage cannot be repaired. In this 'cellular suicide', the death of one cell, or more, may benefit the organism as a whole. For example, in plants the death of the water-conducting xylem cells (tracheids and vessel elements) allows the cells to function more efficiently and so deliver water to the upper parts of a plant.

Aging of the Whole Organism

Organismal senescence is the aging of whole organisms. In general, aging is characterized by the declining ability to respond to stress, increased homeostatic imbalance, and increased risk of aging-associated diseases. Death is the ultimate consequence of aging, though "old age" is not a scientifically recognized cause of death because there is always a specific proximal cause, such as cancer, heart disease, or liver failure. Aging of whole organisms is therefore a complex process that can be defined as "a progressive deterioration of physiological function, an intrinsic age-related process of loss of viability and increase in vulnerability". Differences in maximum life span among species correspond to different "rates of aging". For example, inherited differences in the rate of aging make a mouse elderly at 3 years and a human elderly at 80 years. These genetic differences affect a variety of physiological processes, including the efficiency of DNA repair, antioxidant enzymes, and rates of free radical production.

Senescence of the organism gives rise to the Gompertz–Makeham law of mortality, which says that mortality rate rises rapidly with age.

Some animals, such as some reptiles and fish, age slowly (negligible senescence) and exhibit very long lifespans. Some even exhibit "negative senescence", in which mortality falls with age, in disagreement with the Gompertz–Makeham "law".

Whether replicative senescence (Hayflick limit) plays a causative role in organismal aging is at present an active area of investigation.

Theories of Aging

The process of senescence is complex, and may derive from a variety of different mechanisms and exist for a variety of different reasons. However, senescence is not universal, and scientific evidence suggests that cellular senescence evolved in certain species because it prevents the onset of cancer. In a few simple species, such as Hydra, senescence is negligible and cannot be detected. All such species have no "post-mitotic" cells; they reduce the effect of damaging free radicals by cell division and dilution. Such species are not immortal, however, as they will eventually fall prey to trauma or disease. Moreover, average lifespans can vary greatly within and between species. This suggests that both genetic and environmental factors contribute to aging.

In general, theories that explain senescence have been divided between the programmed and stochastic theories of aging. Programmed theories imply that aging is regulated by biological clocks operating throughout the lifespan. This regulation would depend on changes in gene expression that affect the systems responsible for maintenance, repair, and defense responses. Stochastic theories blame environmental impacts on living organisms that induce cumulative damage at various levels as the cause of aging, examples of which ranging from damage to DNA, damage to tissues and cells by oxygen radicals (widely known as free radicals countered by the even more well-known antioxidants), and cross-linking.

However, aging is seen as a progressive failure of homeodynamics (homeostasis) involving genes for the maintenance and repair, stochastic events leading to molecular damage and molecular heterogeneity, and chance events determining the probability of death. Since complex and interacting systems of maintenance and repair comprise the homeodynamic (old term: homeostasis) space of a biological system, aging is considered to be a progressive shrinkage of homeodynamic space mainly due to increased molecular heterogeneity.

Evolutionary Theories

A gene can be expressed at various life-stages. Therefore, natural selection can support lethal and harmful alleles, if their expression occurs after reproduction. Senescence may be the product of such selection. In addition, aging is believed to have evolved because of the increasingly smaller probability of an organism still being alive at older age, due to predation and accidents, both of which may be random and age-invariant. It is thought that strategies that result in a higher reproductive rate at a young age, but shorter overall lifespan, result in a higher lifetime reproductive success and are therefore favoured by natural selection.

In essence, aging is, therefore, the result of investing resources in reproduction, rather than maintenance of the body (the "Disposable Soma" theory), in light of the fact that accidents, predation, and disease will eventually kill the organism no matter how much energy is devoted to repair of the body. Various other theories of aging exist, and are not necessarily mutually exclusive.

The geneticist J. B. S. Haldane wondered why the dominant mutation that causes Huntington's disease remained in the population, and why natural selection had not eliminated it. The onset of this neurological disease is (on average) at age 45 and is invariably fatal within 10–20 years.

Haldane assumed that, in human prehistory, few survived until age 45. Since few were alive at older ages and their contribution to the next generation was therefore small relative to the large cohorts of younger age groups, the force of selection against such late-acting deleterious mutations was correspondingly small.

However, if a mutation affected younger individuals, selection against it would be strong. Therefore, late-acting deleterious mutations could accumulate in populations over evolutionary time through genetic drift. This principle has been demonstrated experimentally. And it is these later-acting deleterious mutations that are believed to cause—even allow—age-related mortality.

Peter Medawar formalised this observation in his mutation accumulation theory of aging. "The force of natural selection weakens with increasing age—even in a theoretically immortal population, provided only that it is exposed to real hazards of mortality. If a genetic disaster... happens late enough in individual life, its consequences may be completely unimportant".

The 'real hazards of mortality' are, in typical circumstances, predation, disease, and accidents. So, even an immortal population, whose fertility does not decline with time, will have fewer individuals alive in older age groups. This is called 'extrinsic mortality'. Young cohorts, not depleted in numbers yet by extrinsic mortality, contribute far more to the next generation than the few remaining older cohorts, so the force of selection against late-acting deleterious mutations, which affect only these few older individuals, is very weak. The mutations may not be selected against, therefore, and may spread over evolutionary time into the population.

The major testable prediction made by this model is that species that have high extrinsic mortality in nature will age more quickly and have shorter intrinsic lifespans. This is borne out among mammals, the best-studied in terms of life history.

There is a correlation among mammals between body size and lifespan, such that larger species live longer than smaller species in controlled/optimum conditions, but there are notable exceptions. For instance, many bats and rodents are of similar size, yet bats live much longer. For instance, the little brown bat, half the size of a mouse, can live 30 years in the wild. A mouse will live 2–3 years even with optimum conditions. The explanation is that bats have fewer predators, so therefore low extrinsic mortality. Thus, more individuals survive to later ages, so the force of selection against late-acting deleterious mutations is stronger. Fewer late-acting deleterious mutations = slower aging = longer lifespan. Birds are also warm-blooded and are of similar size to many small mammals, yet live often 5–10 times as long. They have fewer predation pressures compared with ground-dwelling mammals. And seabirds, which, in general, have the fewest predators of all birds, live longest.

Also, when examining the body-size vs. lifespan relationship, one will observe that predator mammals tend to have longer lifespans than prey animals in a controlled environment such as a zoo or nature reserve. The explanation for the long lifespans of primates (such as humans, monkeys, and apes) relative to body size is that their intelligence and often sociality helps them avoid becoming prey. Being a predator, being smart, and working together all reduce extrinsic mortality.

Another evolutionary theory of aging was proposed by George C. Williams and involves antagonistic pleiotropy. A single gene may affect multiple traits. Some traits that increase fitness early in life may also have negative effects later in life. But, because many more individuals are alive at young ages than at old ages, even small positive effects early can be strongly selected for, and large negative effects later may be very weakly selected against.

Williams suggested the following example: Perhaps a gene codes for calcium deposition in bones, which promotes juvenile survival and will therefore be favored by natural selection; however, this same gene promotes calcium deposition in the arteries, causing negative effects in old age. Therefore, negative effects in old age may reflect the result of natural selection for pleiotropic genes that are beneficial early in life. In this case, fitness is relatively high when Fisher's reproductive value is high and relatively low when Fisher's reproductive value is low.

Gene Regulation

A number of genetic components of aging have been identified using model organisms, ranging from the simple budding yeast *Saccharomyces cerevisiae* to worms such as *Caenorhabditis elegans* and fruit flies (*Drosophila melanogaster*). Study of these organisms has revealed the presence of at least two conserved aging pathways.

One of these pathways involves the gene *Sir2*, a NAD+-dependent histone deacetylase. In yeast, Sir2 is required for genomic silencing at three loci: The yeast mating loci, the telomeres and the ribosomal DNA (rDNA). In some species of yeast, replicative aging may be partially caused by homologous recombination between rDNA repeats; excision of rDNA repeats results in the formation of extrachromosomal rDNA circles (ERCs). These ERCs replicate and preferentially segregate to the mother cell during cell division, and are believed to result in cellular senescence by titrating away (competing for) essential nuclear factors. ERCs have not been observed in other species (nor even all strains of the same yeast species) of yeast (which also display replicative senescence), and ERCs are not believed to contribute to aging in higher organisms such as humans (they have not been shown to accumulate in mammals in a similar manner to yeast).

Extrachromosomal circular DNA (eccDNA) has been found in worms, flies, and humans. The origin and role of eccDNA in aging, if any, is unknown.

Despite the lack of a connection between circular DNA and aging in higher organisms, extra copies of Sir2 are capable of extending the lifespan of both worms and flies (though, in flies, this finding has not been replicated by other investigators, and the activator of Sir2 resveratrol does not reproducibly increase lifespan in either species).

Whether the Sir2 homologues in higher organisms have any role in lifespan is unclear, but the human SIRT1 protein has been demonstrated to deacetylate p53, Ku70, and the forkhead family of transcription factors. SIRT1 can also regulate acetylates such as CBP/p300, and has been shown to deacetylate specific histone residues.

RAS1 and RAS2 also affect aging in yeast and have a human homologue. RAS2 overexpression has been shown to extend lifespan in yeast. Other genes regulate aging in yeast by increasing the resistance to oxidative stress. Superoxide dismutase, a protein that protects against the effects of mitochondrial free radicals, can extend yeast lifespan in stationary phase when overexpressed.

In higher organisms, aging is likely to be regulated in part through the insulin/IGF-1 pathway. Mutations that affect insulin-like signaling in worms, flies, and the growth hormone/IGF1 axis in mice are associated with extended lifespan.

In yeast, Sir2 activity is regulated by the nicotinamidase PNC1. PNC1 is transcriptionally upregulated under stressful conditions such as caloric restriction, heat shock, and osmotic shock. By converting nicotinamide to niacin, nicotinamide is removed, inhibiting the activity of Sir2. A nicotinamidase found in humans, known as PBEF, may serve a similar function, and a secreted form of PBEF known as visfatin may help to regulate serum insulin levels. It is not known, however, whether these mechanisms also exist in humans, since there are obvious differences in biology between humans and model organisms.

Sir2 activity has been shown to increase under calorie restriction. Due to the lack of available glucose in the cells, more NAD+ is available and can activate Sir2. Resveratrol, a stilbenoid found in the skin of red grapes, was reported to extend the lifespan of yeast, worms, and flies (the lifespan extension in flies and worms have proved irreproducible by independent investigators).

It has been shown to activate Sir2 and therefore mimics the effects of calorie restriction, if one accepts that caloric restriction is indeed dependent on Sir2.

Gene expression is imperfectly controlled, and it is possible that random fluctuations in the expression levels of many genes contribute to the aging process as suggested by a study of such genes in yeast. Individual cells, which are genetically identical, none-the-less can have substantially different responses to outside stimuli, and markedly different lifespans, indicating the epigenetic factors play an important role in gene expression and aging as well as genetic factors.

According to the GenAge database of aging-related genes there are over 700 genes associated with aging in model organisms: 555 in the soil roundworm (*Caenorhabditis elegans*), 87 in the bakers' yeast (*Saccharomyces cerevisiae*), 75 in the fruit fly (*Drosophila melanogaster*) and 68 in the mouse (*Mus musculus*).

Cellular Senescence

As noted above, senescence is not universal, and senescence is not observed in single-celled organisms that reproduce through the process of cellular mitosis.

Moreover, cellular senescence is not observed in several organisms, including perennial plants, sponges, corals, and lobsters. In those species where cellular senescence is observed, cells eventually become post-mitotic when they can no longer replicate themselves through the process of cellular mitosis; i.e., cells experience *replicative senescence*. How and why some cells become post-mitotic in some species has been the subject of much research and speculation, but (as noted above) it is widely believed that cellular senescence evolved as a way to prevent the onset and spread of cancer. Somatic cells that have divided many times will have accumulated DNA mutations and would therefore be in danger of becoming cancerous if cell division continued.

Lately, the role of telomeres in cellular senescence has aroused general interest, especially with a view to the possible genetically adverse effects of cloning. The successive shortening of the chromosomal telomeres with each cell cycle is also believed to limit the number of divisions of the cell, thus contributing to aging. There have, on the other hand, also been reports that cloning could alter the shortening of telomeres. Some cells do not age and are, therefore, described as being "biologically immortal".

It is theorized by some that when it is discovered exactly what allows these cells, whether it be the result of telomere lengthening or not, to divide without limit that it will be possible to genetically alter other cells to have the same capability. It is further theorized that it will eventually be possible to genetically engineer all cells in the human body to have this capability by employing gene therapy and, therefore, stop or reverse aging, effectively making the entire organism potentially immortal.

Cancer cells are usually immortal. In about 85% of tumors, this evasion of cellular senescence is the result of up-activation of their telomerase genes. This simple observation suggests that reactivation of telomerase in healthy individuals could greatly increase their cancer risk. Whether cell senescence plays any role in organismal aging is at present unknown, and is an active area of investigation. Mouse mutants lacking telomerase do not immediately show accelerated aging.

Chemical Damage

One of the earliest aging theories was the *Rate of Living Hypothesis* described by Raymond Pearl in 1928(based on earlier work by Max Rubner), which states that fast basal metabolic rate corresponds to short maximum life span.

While there may be some validity to the idea that for various types of specific damage detailed below that are by-products of metabolism, all other things being equal, a fast metabolism may reduce lifespan, in general this theory does not adequately explain the differences in lifespan either within, or between, species.

Calorically-restricted animals process as much, or more, calories per gram of body mass, as their *ad libitum* fed counterparts, yet exhibit substantially longer lifespans.. Similarly, metabolic rate is a poor predictor of lifespan for birds, bats and other species that, it is presumed, have reduced mortality from predation, and therefore have evolved long lifespans even in the presence of very high metabolic rates. More recently, it was shown that, when modern statistical methods for correcting for the effects of body size and phylogeny are employed, metabolic rate does not correlate with longevity in mammals or birds.

With respect to specific types of chemical damage caused by metabolism, it is suggested that damage to long-lived biopolymers, such as structural proteins or DNA, caused by ubiquitous chemical agents in the body such as oxygen and sugars, are in part responsible for aging. The damage can include breakage of biopolymer chains, cross-linking of biopolymers, or chemical attachment of unnatural substituents (haptens) to biopolymers.

Under normal aerobic conditions, approximately 4% of the oxygen metabolized by mitochondria is converted to superoxide ion, which can subsequently be converted to hydrogen peroxide, hydroxyl radical and eventually other reactive species including other peroxides and singlet oxygen, which can, in turn, generate free radicals capable of damaging structural proteins and DNA. Certain metal ions found in the body, such as copper and iron, may participate in the process. (In Wilson's disease, a hereditary defect that causes the body to retain copper, some of the symptoms resemble accelerated senescence.)

These processes are termed *oxidative damage* and are linked to the benefits of nutritionally derived polyphenol antioxidants. Sugars such as glucose and fructose can react with certain amino acids such as lysine and arginine and certain DNA bases such as guanine to produce sugar adducts, in a process called *glycation*. These adducts can further rearrange to form reactive species, which can then cross-link the structural proteins or DNA to similar biopolymers or other biomolecules such as non-structural proteins. People with diabetes, who have elevated blood sugar, develop senescence-associated disorders much earlier than the general population, but can delay such disorders by rigorous control of their blood sugar levels. There is evidence that sugar damage is linked to oxidant damage in a process termed *glycoxidation*.

Free radicals can damage proteins, lipids or DNA. Glycation mainly damages proteins. Damaged proteins and lipids accumulate in lysosomes as lipofuscin. Chemical damage to structural proteins can lead to loss of function; for example, damage to collagen of blood vessel walls can lead to vessel-wall stiffness and, thus, hypertension, and vessel wall thickening and reactive tissue formation (atherosclerosis);

similar processes in the kidney can lead to renal failure. Damage to enzymes reduces cellular functionality. Lipid peroxidation of the inner mitochondrial membrane reduces the electric potential and the ability to generate energy. It is probably no accident that nearly all of the so-called "accelerated aging diseases" are due to defective DNA repair enzymes.

It is believed that the impact of alcohol on aging can be partly explained by alcohol's activation of the HPA axis, which stimulates glucocorticoid secretion, long-term exposure to which produces symptoms of aging.

Reliability Theory

Reliability theory suggests that biological systems start their adult life with a high load of initial damage. Reliability theory is a general theory about systems failure. It allows researchers to predict the age-related failure kinetics for a system of given architecture (reliability structure) and given reliability of its components.

Reliability theory predicts that even those systems that composed entirely of non-aging elements (with a constant failure rate) will nevertheless deteriorate (fail more often) with age, if these systems are redundant in irreplaceable elements. Aging, therefore, is a direct consequence of systems.

Reliability theory also predicts the late-life mortality deceleration with subsequent leveling-off, as well as the late-life mortality plateaus, as an inevitable consequence of redundancy exhaustion at extreme old ages. The theory explains why mortality rates increase exponentially with age (the Gompertz law) in many species, by taking into account the initial flaws (defects) in newly formed systems.

It also explains why organisms "prefer" to die according to the Gompertz law, while technical devices usually fail according to the Weibull (power) law. Reliability theory allows to specify conditions when organisms die according to the Weibull distribution: Organisms should be relatively free of initial flaws and defects.

The theory makes it possible to find a general failure law applicable to all adult and extreme old ages, where the Gompertz and the Weibull laws are just special cases of this more general failure law. The theory explains why relative differences in mortality rates of compared populations (within a given species) vanish with age (compensation law of mortality), and mortality convergence is observed due to the exhaustion of initial differences in redundancy levels.

Miscellaneous

Recently, a kind of early senescence has been alleged to be a possible unintended outcome of early cloning experiments. The issue was raised in the case of Dolly the sheep, following her death from a contagious lung disease.

The claim that Dolly's early death involved premature senescence has been vigorously contested, and Dolly's creator, Dr. Ian Wilmut has expressed the view that her illness and death were probably unrelated to the fact that she was a clone.

A set of rare hereditary (genetic) disorders, each called progeria, has been known for some time. Sufferers exhibit symptoms resembling accelerated aging, including wrinkled skin. The cause of Hutchinson–Gilford progeria syndrome was reported in the journal *Nature* in May 2003. This report suggests that DNA damage, not oxidative stress, is the cause of this form of accelerated aging.

Transcription

Transcription is the process of creating a complementary RNA copy of a sequence of DNA. Both RNA and DNA are nucleic acids, which use base pairs of nucleotides as a complementary language that can be converted back and forth from DNA to RNA by the action of the correct enzymes. During transcription, a DNA sequence is read by RNA polymerase, which produces a complementary, antiparallel RNA strand. As opposed to DNA replication, transcription results in an RNA complement that includes uracil (U) in all instances where thymine (T) would have occurred in a DNA complement.

Transcription can be explained easily in 4 or 5 simple steps, each moving like a wave along the DNA.

1. RNA polymerase unwinds/"unzips" the DNA by breaking the hydrogen bonds between complimentary nucleotides.
2. RNA nucleotides are paired with complementary DNA bases.
3. RNA sugar-phosphate backbone forms with assistance from RNA polymerase.
4. Hydrogen bonds of the untwisted RNA+DNA helix break, freeing the newly synthesized RNA strand.
5. If the cell has a nucleus, the RNA is further processed and then moves through the small nuclear pores to the cytoplasm.

Transcription is the first step leading to gene expression. The stretch of DNA transcribed into an RNA molecule is called a *transcription unit* and encodes at least one gene. If the gene transcribed encodes a protein, the result of transcription is messenger RNA (mRNA), which will then be used to create that protein via the process of translation. Alternatively, the transcribed gene may encode for either ribosomal RNA (rRNA) or transfer RNA (tRNA), other components of the protein-assembly process, or other ribozymes.

A DNA transcription unit encoding for a protein contains not only the sequence that will eventually be directly translated into the protein (the *coding sequence*) but also *regulatory sequences* that direct and regulate the synthesis of that protein. The regulatory sequence before (upstream from) the coding sequence is called the five prime untranslated region (5'UTR), and the sequence following (downstream from) the coding sequence is called the three prime untranslated region (3'UTR).

Transcription has some proofreading mechanisms, but they are fewer and less effective than the controls for copying DNA; therefore, transcription has a lower copying fidelity than DNA replication.

As in DNA replication, DNA is read from 3' → 5' during transcription. Meanwhile, the complementary RNA is created from the 5' → 3' direction. This means its 5' end is created first in base pairing. Although DNA is arranged as two antiparallel strands in a double helix, only one of the two DNA strands, called the template strand, is used for transcription. This is because RNA is only single-stranded, as opposed to double-stranded DNA. The other DNA strand is called the coding strand, because its sequence is the same as the newly created RNA transcript (except for the substitution of uracil for thymine). The use of only the 3' → 5' strand eliminates the need for the Okazaki fragments seen in DNA replication.

Transcription is divided into 5 stages: *pre-initiation, initiation, promoter clearance, elongation* and *termination.*

Major Steps

Pre-initiation

In eukaryotes, RNA polymerase, and therefore the initiation of transcription, requires the presence of a core promoter sequence in the DNA. Promoters are regions of DNA that promote transcription and, in eukaryotes, are found at -30, -75, and -90 base pairs upstream from the start site of transcription. Core promoters are sequences within the promoter that are essential for transcription initiation. RNA polymerase is able to bind to core promoters in the presence of various specific transcription factors. The most common type of core promoter in eukaryotes is a short DNA sequence known as a TATA box, found 25-30 base pairs upstream from the start site of transcription. The TATA box, as a core promoter, is the binding site for a transcription factor known as TATA-binding protein (TBP), which is itself a subunit of another transcription factor, called Transcription Factor II D (TFIID).

After TFIID binds to the TATA box via the TBP, five more transcription factors and RNA polymerase combine around the TATA box in a series of stages to form a preinitiation complex. One transcription factor, DNA helicase, has helicase activity and so is involved in the separating of opposing strands of double-stranded DNA to provide access to a single-stranded DNA template.

However, only a low, or basal, rate of transcription is driven by the preinitiation complex alone. Other proteins known as activators and repressors, along with any associated coactivators or corepressors, are responsible for modulating transcription rate. Thus, preinitiation complex contains:

1. Core Promoter Sequence
2. Transcription Factors
3. DNA Helicase
4. RNA Polymerase
5. Activators and Repressors

The transcription preinitiation in archaea is, in essence, homologous to that of eukaryotes, but is much less complex. The archaeal preinitiation complex assembles at a TATA-box binding site; however, in archaea, this complex is composed of only RNA polymerase II, TBP, and TFB (the archaeal homologue of eukaryotic transcription factor II B (TFIIB)).

Initiation

In bacteria, transcription begins with the binding of RNA polymerase to the promoter in DNA. RNA polymerase is a core enzyme consisting of five subunits: 2 α subunits, 1 β subunit, 1 β' subunit, and 1 ω subunit. At the start of initiation, the core enzyme is associated with a sigma factor that aids in finding the appropriate -35 and -10 base pairs downstream of promoter sequences. Transcription initiation is more complex in eukaryotes.

Eukaryotic RNA polymerase does not directly recognize the core promoter sequences. Instead, a collection of proteins called transcription factors mediate the binding of RNA polymerase and the initiation of transcription. Only after certain transcription factors are attached to the promoter does the RNA polymerase bind to it. The completed assembly of transcription factors and RNA polymerase bind to the promoter, forming a transcription initiation complex. Transcription in the archaea domain is similar to transcription in eukaryotes.

Promoter Clearance

After the first bond is syntheized, the RNA polymerase must clear the promoter. During this time there is a tendency to release the RNA transcript and produce truncated transcripts. This is called *abortive initiation* and is common for both eukaryotes and prokaryotes. Abortive initiation continues to occur until the σ factor rearranges, resulting in the transcription elongation complex (which gives a 35 bp moving footprint). The σ factor is released before 80 nucleotides of mRNA are synthesized. Once the transcript reaches approximately 23 nucleotides, it no longer slips and elongation can occur. This, like most of the remainder of transcription, is an energy-dependent process, consuming adenosine triphosphate (ATP).

Promoter clearance coincides with phosphorylation of serine 5 on the carboxy terminal domain of RNA Pol in eukaryotes, which is phosphorylated by TFIIH.

Elongation

One strand of the DNA, the *template strand* (or noncoding strand), is used as a template for RNA synthesis. As transcription proceeds, RNA polymerase traverses the template strand and uses base pairing complementarity with the DNA template to create an RNA copy.

Although RNA polymerase traverses the template strand from 3' → 5', the coding (non-template) strand and newly-formed RNA can also be used as reference points, so transcription can be described as occurring 5' → 3'. This produces an RNA molecule from 5' → 3', an exact copy of the coding strand (except that thymines are replaced with uracils, and the nucleotides are composed of a ribose (5-carbon) sugar where DNA has deoxyribose (one less oxygen atom) in its sugar-phosphate backbone).

Unlike DNA replication, mRNA transcription can involve multiple RNA polymerases on a single DNA template and multiple rounds of transcription (amplification of particular mRNA), so many mRNA molecules can be rapidly produced from a single copy of a gene.

Elongation also involves a proofreading mechanism that can replace incorrectly incorporated bases. In eukaryotes, this may correspond with short pauses during transcription that allow appropriate RNA editing factors to bind. These pauses may be intrinsic to the RNA polymerase or due to chromatin structure.

Termination

Bacteria use two different strategies for transcription termination. In Rho-independent transcription termination, RNA transcription stops when the newly synthesized RNA molecule forms a G-C-rich hairpin loop followed by a run of Us.

When the hairpin forms, the mechanical stress breaks the weak rU-dA bonds, now filling the DNA-RNA hybrid. This pulls the poly-U transcript out of the active site of the RNA polymerase, in effect, terminating transcription. In the "Rho-dependent" type of termination, a protein factor called "Rho" destabilizes the interaction between the template and the mRNA, thus releasing the newly synthesized mRNA from the elongation complex.

Transcription termination in eukaryotes is less understood but involves cleavage of the new transcript followed by template-independent addition of *A*s at its new 3' end, in a process called polyadenylation.

Measuring and Detecting Transcription

Transcription can be measured and detected in a variety of ways:

- Nuclear Run-on assay: measures the relative abundance of newly formed transcripts
- RNase protection assay and ChIP-Chip of RNAP: detect active transcription sites
- RT-PCR: measures the absolute abundance of total or nuclear RNA levels, which may however differ from transcription rates
- DNA microarrays: measures the relative abundance of the global total or nuclear RNA levels; however, these may differ from transcription rates
- In situ hybridization: detects the presence of a transcript
- MS2 tagging: by incorporating RNA stem loops, such as MS2, into a gene, these become incorporated into newly synthesized RNA. The stem loops can then be detected using a fusion of GFP and the MS2 coat protein, which has a high affinity, sequence-specific interaction with the MS2 stem loops. The recruitment of GFP to the site of transcription is visualised as a single fluorescent spot. This remarkable new approach has

revealed that transcription occurs in discontinuous bursts, or pulses. With the notable exception of in situ techniques, most other methods provide cell population averages, and are not capable of detecting this fundamental property of genes.

- Northern blot: the traditional method, and until the advent of RNA-Seq, the most quantitative
- RNA-Seq: applies next-generation sequencing techniques to sequence whole transcriptomes, which allows the measurement of relative abundance of RNA, as well as the detection of additional variations such as fusion genes, post-translational edits and novel splice sites.

Transcription Factories

Active transcription units are clustered in the nucleus, in discrete sites called transcription factories or euchromatin. Such sites can be visualized by allowing engaged polymerases to extend their transcripts in tagged precursors (Br-UTP or Br-U) and immuno-labelling the tagged nascent RNA.

Transcription factories can also be localized using fluorescence in situ hybridization or marked by antibodies directed against polymerases.

There are ~10,000 factories in the nucleoplasm of a HeLa cell, among which are ~8,000 polymerase II factories and ~2,000 polymerase III factories. Each polymerase II factory contains ~8 polymerases.

As most active transcription units are associated with only one polymerase, each factory usually contains ~8 different transcription units. These units might be associated through promoters and/or enhancers, with loops forming a 'cloud' around the factor.

History

A molecule that allows the genetic material to be realized as a protein was first hypothesized by François Jacob and Jacques Monod. RNA synthesis by RNA polymerase was established *in vitro* by several laboratories by 1965; however, the RNA synthesized by these enzymes had properties that suggested the existence of an additional factor needed to terminate transcription correctly.

In 1972, Walter Fiers became the first person to actually prove the existence of the terminating enzyme.

Roger D. Kornberg won the 2006 Nobel Prize in Chemistry "for his studies of the molecular basis of eukaryotic transcription".

Reverse Transcription

Some viruses (such as HIV, the cause of AIDS), have the ability to transcribe RNA into DNA. HIV has an RNA genome that is duplicated into DNA. The resulting DNA can be merged with the DNA genome of the host cell. The main enzyme responsible for synthesis of DNA from an RNA template is called reverse transcriptase.

In the case of HIV, reverse transcriptase is responsible for synthesizing a complementary DNA strand (cDNA) to the viral RNA genome. An associated enzyme, ribonuclease H, digests the RNA strand, and reverse transcriptase synthesises a complementary strand of DNA to form a double helix DNA structure.

This cDNA is integrated into the host cell's genome via another enzyme (integrase) causing the host cell to generate viral proteins that reassemble into new viral particles. In HIV, subsequent to this, the host cell undergoes programmed cell death, apoptosis of T cells. However, in other retroviruses, the host cell remains intact as the virus buds out of the cell.

Some eukaryotic cells contain an enzyme with reverse transcription activity called telomerase. Telomerase is a reverse transcriptase that lengthens the ends of linear chromosomes. Telomerase carries an RNA template from which it synthesizes DNA repeating sequence, or "junk" DNA. This repeated sequence of DNA is important because, every time a linear chromosome is duplicated, it is shortened in length. With "junk" DNA at the ends of chromosomes, the shortening eliminates some of the non-essential, repeated sequence rather than the protein-encoding DNA sequence farther away from the chromosome end.

Telomerase is often activated in cancer cells to enable cancer cells to duplicate their genomes indefinitely without losing important protein-coding DNA sequence. Activation of telomerase could be part of the process that allows cancer cells to become *immortal*. However, the true *in vivo* significance of telomerase has still not been empirically proven.

Ribonucleic Acid

Ribonucleic acid (RNA) is one of the three major macromolecules (along with DNA and proteins) that are essential for all known forms of life. Like DNA, RNA is made up of a long chain of components called nucleotides. Each nucleotide consists of a nucleobase (sometimes called a nitrogenous base), a ribose sugar, and a phosphate group. The sequence of nucleotides allows RNA to encode genetic information. For example, some viruses use RNA instead of DNA as their genetic material, and all organisms use messenger RNA (mRNA) to carry the genetic information that directs the synthesis of proteins.

Like proteins, some RNA molecules play an active role in cells by catalyzing biological reactions, controlling gene expression, or sensing and communicating responses to cellular signals. One of these active processes is protein synthesis, a universal function whereby mRNA molecules direct the assembly of proteins on ribosomes. This process uses transfer RNA (tRNA) molecules to deliver amino acids to the ribosome, where ribosomal RNA (rRNA) links amino acids together to form proteins.

The chemical structure of RNA is very similar to that of DNA, with two differences—(a) RNA contains the sugar ribose while DNA contains the slightly different sugar deoxyribose (a type of ribose that lacks one oxygen atom), and (b) RNA has the nucleobase uracil while DNA contains thymine (uracil and thymine have similar base-pairing properties). Unlike DNA, most RNA molecules are single-stranded. Single-stranded RNA molecules adopt very complex three-dimensional structures, since they are not restricted to the repetitive double-helical form of double-stranded DNA. RNA is made within living cells by RNA polymerases, enzymes that act to copy a DNA or RNA template into a new RNA strand through processes known as transcription or RNA replication, respectively.

Comparison with DNA

RNA and DNA are both nucleic acids, but differ in three main ways. First, unlike DNA, which is, in general, double-stranded, RNA is a single-stranded molecule in many of its biological roles and has a much shorter chain of nucleotides. Second, while DNA contains *deoxyribose*, RNA contains *ribose* (in deoxyribose there is no hydroxyl group attached to the pentose ring in the 2' position). These hydroxyl groups make RNA less stable than DNA because it is more prone to hydrolysis. Third, the complementary base to adenine is not thymine, as it is in DNA, but rather uracil, which is an unmethylated form of thymine.

Like DNA, most biologically active RNAs, including mRNA, tRNA, rRNA, snRNAs, and other non-coding RNAs, contain self-complementary sequences that allow parts of the RNA to fold and pair with itself to form double helices. Structural analysis of these RNAs has revealed that they are highly structured. Unlike DNA, their structures do not consist of long double helices but rather collections of short helices packed together into structures akin to proteins. In this fashion, RNAs can achieve chemical catalysis, like enzymes. For instance, determination of the structure of the ribosome—an enzyme that catalyzes peptide bond formation—revealed that its active site is composed entirely of RNA.

Structure

Each nucleotide in RNA contains a ribose sugar, with carbons numbered 1' through 5'. A base is attached to the 1' position, in general, adenine (A), cytosine (C), guanine (G), or uracil (U). Adenine and guanine are purines, cytosine, and uracil are pyrimidines. A phosphate group is attached to the 3' position of one ribose and the 5' position of the next. The phosphate groups have a negative charge each at physiological pH, making RNA a charged molecule (polyanion).

The bases may form hydrogen bonds between cytosine and guanine, between adenine and uracil and between guanine and uracil. However, other interactions are possible, such as a group of adenine bases binding to each other in a bulge, or the GNRA tetraloop that has a guanine–adenine base-pair.

An important structural feature of RNA that distinguishes it from DNA is the presence of a hydroxyl group at the 2' position of the ribose sugar. The presence of this functional group causes the helix to adopt the A-form geometry rather than the B-form most commonly observed in DNA. This results in a very deep and narrow major groove and a shallow and wide minor groove. A second consequence of the presence of the 2'-hydroxyl group is that in conformationally flexible regions of an RNA molecule (that is, not involved in formation of a double helix), it can chemically attack the adjacent phosphodiester bond to cleave the backbone.

RNA is transcribed with only four bases (adenine, cytosine, guanine and uracil), but these bases and attached sugars can be modified in numerous ways as the RNAs mature. Pseudouridine (Ø), in which the linkage between uracil and ribose is changed from a C–N bond to a C–C bond, and ribothymidine (T) are found in various places (the most notable ones being in the TØC loop of tRNA). Another notable modified base is hypoxanthine, a deaminated adenine base whose nucleoside is called inosine (I). Inosine plays a key role in the wobble hypothesis of the genetic code.

There are nearly 100 other naturally occurring modified nucleosides, of which pseudouridine and nucleosides with 2'-Omethylribose are the most common. The specific roles of many of these modifications in RNA are not fully understood. However, it is notable that, in ribosomal RNA, many of the post-transcriptional modifications occur in highly functional regions, such as the peptidyl transferase center and the subunit interface, implying that they are important for normal function.

The functional form of single stranded RNA molecules, just like proteins, frequently requires a specific tertiary structure.

The scaffold for this structure is provided by secondary structural elements that are hydrogen bonds within the molecule. This leads to several recognizable "domains" of secondary structure like hairpin loops, bulges, and internal loops. Since RNA is charged, metal ions such as Mg^{2+} are needed to stabilise many secondary and tertiary structures.

Synthesis

Synthesis of RNA is usually catalyzed by an enzyme—RNA polymerase—using DNA as a template, a process known as transcription. Initiation of transcription begins with the binding of the enzyme to a promoter sequence in the DNA (usually found "upstream" of a gene).

The DNA double helix is unwound by the helicase activity of the enzyme. The enzyme then progresses along the template strand in the 3' to 5' direction, synthesizing a complementary RNA molecule with elongation occurring in the 5' to 3' direction. The DNA sequence also dictates where termination of RNA synthesis will occur.

RNAs are often modified by enzymes after transcription. For example, a poly(A) tail and a 5' cap are added to eukaryotic pre-mRNA and introns are removed by the spliceosome.

There are also a number of RNA-dependent RNA polymerases that use RNA as their template for synthesis of a new strand of RNA. For instance, a number of RNA viruses (such as poliovirus) use this type of enzyme to replicate their genetic material. Also, RNA-dependent RNA polymerase is part of the RNA interference pathway in many organisms.

Types of RNA

Messenger RNA (mRNA) is the RNA that carries information from DNA to the ribosome, the sites of protein synthesis (translation) in the cell. The coding sequence of the mRNA determines the amino acid sequence in the protein that is produced. Many RNAs do not code for protein however (about 97% of the transcriptional output is non-protein-coding in eukaryotes).

These so-called non-coding RNAs ("ncRNA") can be encoded by their own genes (RNA genes), but can also derive from mRNA introns. The most prominent examples of non-coding RNAs are transfer RNA (tRNA) and ribosomal RNA (rRNA), both of which are involved in the process of translation.

There are also non-coding RNAs involved in gene regulation, RNA processing and other roles. Certain RNAs are able to catalyse chemical reactions such as cutting and ligating other RNA molecules, and the catalysis of peptide bond formation in the ribosome; these are known as ribozymes.

In Translation

Messenger RNA (mRNA) carries information about a protein sequence to the ribosomes, the protein synthesis factories in the cell. It is coded so that every three nucleotides (a codon) correspond to one amino acid. In eukaryotic cells, once precursor mRNA (pre-mRNA) has been transcribed from DNA, it is processed to mature mRNA.

This removes its introns—non-coding sections of the pre-mRNA. The mRNA is then exported from the nucleus to the cytoplasm, where it is bound to ribosomes and translated into its corresponding protein form with the help of tRNA.

In prokaryotic cells, which do not have nucleus and cytoplasm compartments, mRNA can bind to ribosomes while it is being transcribed from DNA. After a certain amount of time the message degrades into its component nucleotides with the assistance of ribonucleases.

Transfer RNA (tRNA) is a small RNA chain of about 80 nucleotides that transfers a specific amino acid to a growing polypeptide chain at the ribosomal site of protein synthesis during translation. It has sites for amino acid attachment and an anticodon region for codon recognition that binds to a specific sequence on the messenger RNA chain through hydrogen bonding. Ribosomal RNA (rRNA) is the catalytic component of the ribosomes. Eukaryotic ribosomes contain four different rRNA molecules: 18S, 5.8S, 28S and 5S rRNA.

Three of the rRNA molecules are synthesized in the nucleolus, and one is synthesized elsewhere. In the cytoplasm, ribosomal RNA and protein combine to form a nucleoprotein called a ribosome. The ribosome binds mRNA and carries out protein synthesis. Several ribosomes may be attached to a single mRNA at any time. rRNA is extremely abundant and makes up 80% of the 10 mg/ml RNA found in a typical eukaryotic cytoplasm.

Transfer-messenger RNA (tmRNA) is found in many bacteria and plastids. It tags proteins encoded by mRNAs that lack stop codons for degradation and prevents the ribosome from stalling.

Regulatory RNAs

Several types of RNA can downregulate gene expression by being complementary to a part of an mRNA or a gene's DNA. MicroRNAs (miRNA; 21-22 nt) are found in eukaryotes and act through RNA interference (RNAi), where an effector complex of miRNA and enzymes can break down mRNA to which the miRNA is complementary, block the mRNA from being translated, or accelerate its degradation. While small interfering RNAs (siRNA; 20-25 nt) are often produced by breakdown of viral RNA, there are also endogenous sources of siRNAs.

siRNAs act through RNA interference in a fashion similar to miRNAs. Some miRNAs and siRNAs can cause genes they target to be methylated, thereby decreasing or increasing transcription of those genes. Animals have Piwi-interacting RNAs (piRNA; 29-30 nt) which are active in germline cells and are thought to be a defense against transposons and play a role in gametogenesis. Many prokaryotes have CRISPR RNAs, a regulatory system similar to RNA interference. Antisense RNAs are widespread; most downregulate a gene, but a few are activators of transcription. One way antisense RNA can act is by binding to an mRNA, forming double-stranded RNA that is enzymatically degraded.

There are many long noncoding RNAs that regulate genes in eukaryotes, one such RNA is Xist, which coats one X chromosome in female mammals and inactivates it. An mRNA may contain regulatory elements itself, such as riboswitches, in the 5' untranslated region or 3' untranslated region; these cis-regulatory elements regulate the activity of that mRNA. The untranslated regions can also contain elements that regulate other genes.

In RNA Processing

Many RNAs are involved in modifying other RNAs. Introns are spliced out of pre-mRNA by spliceosomes, which contain several small nuclear RNAs (snRNA), or the introns can be ribozymes that are spliced by themselves. RNA can also be altered by having its nucleotides modified to other nucleotides than A, C, G and U.

In eukaryotes, modifications of RNA nucleotides are generally directed by small nucleolar RNAs (snoRNA; 60-300 nt), found in the nucleolus and cajal bodies. snoRNAs associate with enzymes and guide them to a spot on an RNA by basepairing to that RNA. These enzymes then perform the nucleotide modification. rRNAs and tRNAs are extensively modified, but snRNAs and mRNAs can also be the target of base modification.

RNA Genomes

Like DNA, RNA can carry genetic information. RNA viruses have genomes composed of RNA, and a variety of proteins encoded by that genome. The viral genome is replicated by some of those proteins, while other proteins protect the genome as the virus particle moves to a new host cell. Viroids are another group of pathogens, but they consist only of RNA, do not encode any protein and are replicated by a host plant cell's polymerase.

In Reverse Transcription

Reverse transcribing viruses replicate their genomes by reverse transcribing DNA copies from their RNA; these DNA copies are then transcribed to new RNA. Retrotransposons also spread by copying DNA and RNA from one another, and telomerase contains an RNA that is used as template for building the ends of eukaryotic chromosomes.

Double-stranded RNA

Double-stranded RNA (dsRNA) is RNA with two complementary strands, similar to the DNA found in all cells. dsRNA forms the genetic material of some viruses (double-stranded RNA viruses). Double-stranded RNA such as viral RNA or siRNA can trigger RNA interference in eukaryotes, as well as interferon response in vertebrates.

Key Discoveries in RNA Biology

Research on RNA has led to many important biological discoveries and numerous Nobel Prizes. Nucleic acids were discovered in 1868 by Friedrich Miescher, who called the material 'nuclein' since it was found in the nucleus. It was later discovered that prokaryotic cells, which do not have a nucleus, also contain nucleic acids. The role of RNA in protein synthesis was suspected already in 1939. Severo Ochoa won the 1959 Nobel Prize in Medicine (shared with Arthur Kornberg) after he discovered an enzyme that can synthesize RNA in the laboratory. Ironically, the enzyme discovered by Ochoa (polynucleotide phosphorylase) was later shown to be responsible for RNA degradation, not RNA synthesis.

The sequence of the 77 nucleotides of a yeast tRNA was found by Robert W. Holley in 1965, winning Holley the 1968 Nobel Prize in

Medicine (shared with Har Gobind Khorana and Marshall Nirenberg). In 1967, Carl Woese hypothesized that RNA might be catalytic and suggested that the earliest forms of life (self-replicating molecules) could have relied on RNA both to carry genetic information and to catalyze biochemical reactions—an RNA world.

During the early 1970s, retroviruses and reverse transcriptase were discovered, showing for the first time that enzymes could copy RNA into DNA (the opposite of the usual route for transmission of genetic information).

For this work, David Baltimore, Renato Dulbecco and Howard Temin were awarded a Nobel Prize in 1975. In 1976, Walter Fiers and his team determined the first complete nucleotide sequence of an RNA virus genome, that of bacteriophage MS2.

In 1977, introns and RNA splicing were discovered in both mammalian viruses and in cellular genes, resulting in a 1993 Nobel to Philip Sharp and Richard Roberts. Catalytic RNA molecules (ribozymes) were discovered in the early 1980s, leading to a 1989 Nobel award to Thomas Cech and Sidney Altman. In 1990 it was found in petunia that introduced genes can silence similar genes of the plant's own, now known to be a result of RNA interference.

At about the same time, 22 nt long RNAs, now called microRNAs, were found to have a role in the development of *C. elegans*. Studies on RNA interference gleaned a Nobel Prize for Andrew Fire and Craig Mello in 2006, and another Nobel was awarded for studies on transcription of RNA to Roger Kornberg in the same year. The discovery of gene regulatory RNAs has led to attempts to develop drugs made of RNA, such as siRNA, to silence genes.

Transcription and Processing of RNA

Gene Expression: Transcription

The majority of genes are expressed as the proteins they encode. The process occurs in two steps:

- Transcription = DNA $\rightarrow$ RNA
- Translation = RNA $\rightarrow$ protein.

Taken together, they make up the "central dogma" of biology: DNA $\rightarrow$ RNA $\rightarrow$ protein.

Gene Transcription: DNA $\rightarrow$ RNA

DNA serves as the template for the synthesis of RNA much as it does for its own replication.

The Steps

- Some 50 different protein transcription factors bind to promoter sites, usually on the 52 side of the gene to be transcribed.
- An enzyme, an RNA polymerase, binds to the complex of transcription factors.
- Working together, they open the DNA double helix.
- The RNA polymerase proceeds to "read" one strand moving in its 32 → 52 direction.
- In eukaryotes, this requires — at least for protein-encoding genes — that the nucleosomes in front of the advancing RNA polymerase (RNAP II) be removed. A complex of proteins is responsible for this. The same complex replaces the nucleosomes after the DNA has been transcribed and RNAP II has moved on.
- As the RNA polymerase travels along the DNA strand, it assembles ribonucleotides (supplied as triphosphates, e.g., ATP) into a strand of RNA.
- Each ribonucleotide is inserted into the growing RNA strand following the rules of base pairing. Thus for each C encountered on the DNA strand, a G is inserted in the RNA; for each G, a C; and for each T, an A. However, each A on the DNA guides the insertion of the pyrimidine uracil (U, from uridine triphosphate, UTP). There is no T in RNA.

Quality control: Occasionally RNA polymerase will select and insert an incorrect, mismatched, ribonucleotide. When this occurs in bacteria (and perhaps in all organisms), the enzyme backs up, removes the incorrect nucleotide (and the one preceding it) and tries again. (Described by Zenkin *et al.*, in the 28 July 2006 issue of Science.)

- Synthesis of the RNA proceeds in the 52 → 32 direction.
- As each nucleoside triphosphate is brought in to add to the 32 end of the growing strand, the two terminal phosphates are removed.
- When transcription is complete, the transcript is released from the polymerase and, shortly thereafter, the polymerase is released from the DNA.

Note that at any place in a DNA molecule, either strand may be serving as the template; that is, some genes "run" one way, some the other (and in a few remarkable cases, the same segment of double helix contains genetic information on both strands!). In all cases, however, RNA polymerase transcribes the DNA strand in its 32 → 52 direction. A report in the 4 January 2001 issue of Nature shows that RNA polymerase actually tracks around the double helix of DNA. In vitro, at least, when RNA polymerase is immobilized, it spins the DNA molecule around and around as it moves along the molecule. Whether it is the polymerase or the DNA that does the spinning in vivo remains to be determined.

Types of RNA

Sedimentation pattern produced by high-speed centrifugation of RNA extracted from the precursors of rabbit red blood cells. The discrete bands represent particular classes of RNA.

The transfer RNAs band at about 4S. The ribosomal RNAs of eukaryotes sediment at 5S, 5.8S, 18S, and 28S. (The larger the sedimentation unit, S, the larger the molecule — but not proportionally.) The RNA forming the band at 9S is the messenger RNA for the synthesis of hemoglobin, the major protein synthesized in these cells. In most types of cells, the messenger RNAs are extremely heterogenous, with small amounts distributed from 6S to 25S.

Several types of RNA are synthesized in the nucleus of eukaryotic cells. Of particular interest are:

- messenger RNA (mRNA). This will later be translated into a polypeptide.
- ribosomal RNA (rRNA). This will be used in the building of ribosomes: machinery for synthesizing proteins by translating mRNA.
- transfer RNA (tRNA). RNA molecules that carry amino acids to the growing polypeptide.
- small nuclear RNA (snRNA). DNA transcription of the genes for mRNA, rRNA, and tRNA produces large precursor molecules ("primary transcripts") that must be processed within the nucleus to produce the functional molecules for export to the cytosol. Some of these processing steps are mediated by snRNAs.
- small nucleolar RNA (snoRNA). These RNAs within the nucleolus have several functions (described below).
- microRNA (miRNA). These are tiny (~22 nucleotides) RNA molecules that appear to regulate the expression of messenger RNA (mRNA) molecules. [Discussion]
- XIST RNA. This inactivates one of the two X chromosomes in female vertebrates. [Discussion].

Messenger RNA (mRNA)

Messenger RNA comes in a wide range of sizes reflecting the size of the polypeptide it encodes. Most cells produce small amounts of thousands of different mRNA molecules, each to be translated into a peptide needed by the cell.

Many mRNAs are common to most cells, encoding "housekeeping" proteins needed by all cells (e.g. the enzymes of glycolysis). Other mRNAs are specific for only certain types of cells. These encode proteins needed for the function of that particular cell (e.g., the mRNA for hemoglobin in the precursors of red blood cells).

Ribosomal RNA (rRNA)

There are 4 kinds. In eukaryotes, these are:

- 18S rRNA. One of these molecules, along with some 30 different protein molecules, is used to make the small subunit of the ribosome.
- 28S, 5.8S, and 5S rRNA. One each of these molecules, along with some 45 different proteins, are used to make the large subunit of the ribosome.

The S number given each type of rRNA reflects the rate at which the molecules sediment in the ultracentrifuge. The larger the number, the larger the molecule (but not proportionally).

The 28S, 18S, and 5.8S molecules are produced by the processing of a single primary transcript from a cluster of identical copies of a single gene. The 5S molecules are produced from a different cluster of identical genes.

Transfer RNA (tRNA)

There are some 32 different kinds of tRNA in a typical eukaryotic cell.

- Each is the product of a separate gene.
- They are small (~4S), containing 73-93 nucleotides.
- Many of the bases in the chain pair with each other forming sections of double helix.
- The unpaired regions form 3 loops.
- Each kind of tRNA carries (at its 32 end) one of the 20 amino acids (thus most amino acids have more than one tRNA responsible for them).
- At one loop, 3 unpaired bases form an anticodon.

- Base pairing between the anticodon and the complementary codon on a mRNA molecule brings the correct amino acid into the growing polypeptide chain.

Small Nuclear RNA (snRNA)

Approximately a dozen different genes for snRNAs, each present in multiple copies, have been identified. The snRNAs have various roles in the processing of the other classes of RNA. For example, several snRNAs are part of the spliceosomes that participate in converting pre-mRNA into mRNA by excising the introns and splicing the exons.

Small Nucleolar RNA (snoRNA)

As the name suggests, these small (60–300 nucleotides) RNAs are found in the nucleolus where they are responsible for several functions:

- Some participate in making ribosomes by helping to cut up the large RNA precursor of the 28S, 18S, and 5.8S molecules.
- Others chemically modify many of the nucleotides in rRNA, tRNA, and snRNA molecules, e.g., by adding methyl groups to ribose.
- Some have been implicated in the alternative splicing of pre-mRNA to different forms of mature mRNA.
- One snoRNA serves as the template for the synthesis of telomeres.

In vertebrates, the snoRNAs are made from introns removed during RNA processing.

Noncoding RNA

Only messenger RNA encodes polypeptides. All the other classes of RNA, including types not mentioned here, are thus called noncoding RNA. Much remains to be learned about the function(s) of some of them. But, taken together, noncoding RNAs probably account for two-thirds of the transcription going on in the nucleus.

The RNA Polymerases

The RNA polymerases are huge multi-subunit protein complexes. Three kinds are found in eukaryotes.

- RNA polymerase I (Pol I). It transcribes the rRNA genes for the precursor of the 28S, 18S, and 5.8S molecules (and is the busiest of the RNA polymerases).
- RNA polymerase II (Pol II; also known as RNAP II). It transcribes protein-encoding genes into mRNA (and also the snRNA genes).

- RNA polymerase III (Pol III). It transcribes the 5S rRNA genes and all the tRNA genes.

Two remarkable reports in the 8 June 2001 issue of Science show the structure of Pol II and reveal many details of how it uses the antisense strand of DNA to synthesize a strand of mRNA.

RNA Processing: Pre-mRNA → mRNA

All the primary transcripts produced in the nucleus must undergo processing steps to produce functional RNA molecules for export to the cytosol. We shall confine ourselves to a view of the steps as they occur in the processing of pre-mRNA to mRNA. Most eukaryotic genes are split into segments. In decoding the open reading frame of a gene for a known protein, one usually encounters periodic stretches of DNA calling for amino acids that do not occur in the actual protein product of that gene. Such stretches of DNA, which get transcribed into RNA but not translated into protein, are called introns. Those stretches of DNA that do code for amino acids in the protein are called exons. Examples:

- The gene for one type of collagen found in chickens is split into 52 separate exons.
- The gene for dystrophin, which is mutated in boys with muscular dystrophy, has 79 exons.
- Even the genes for rRNA and tRNA are split by introns.
- The human genome is estimated to contain some 180,000 exons. With a current estimate of 21,000 genes, the average exon content of our genes is about 9.

In general, introns tend to be much longer than exons. An average eukaryotic exon is only 140 nucleotides long, but one human intron stretches for 480,000 nucleotides!

Removal of the introns — and splicing the exons together — are among the essential steps in synthesizing mRNA.

The Steps of RNA Processing

- Synthesis of the cap. This is a modified guanine (G) which is attached to the 52 end of the pre-mRNA as it emerges from RNA polymerase II (RNAP II). The cap
  - — protects the RNA from being degraded by enzymes that degrade RNA from the 52 end;
  - — serves as an assembly point for the proteins needed to recruit the small subunit of the ribosome to begin translation.

- Step-by-step removal of introns present in the pre-mRNA and splicing of the remaining exons. This step takes place as the pre-mRNA continues to emerge from RNAP II.
- Synthesis of the poly(A) tail. This is a stretch of adenine (A) nucleotides. When a special poly(A) attachment site in the pre-mRNA emerges from RNAP II, the transcript is cut there, and the poly(A) tail is attached to the exposed 32 end. This completes the mRNA molecule, which is now ready for export to the cytosol. (The remainder of the tanscript is degraded, and the RNA polymerase leaves the DNA).

The cutting and splicing of mRNA must be done with great precision. If even one nucleotide is left over from an intron or one is removed from an exon, the reading frame from that point on will be shifted, producing new codons specifying a totally different sequence of amino acids from that point to the end of the molecule (which often ends prematurely anyway when the shifted reading frame generates a STOP codon). The removal of introns and splicing of exons is done by spliceosomes. These are a complexes of 5 snRNA molecules and some 145 different proteins. The introns in most pre-mRNAs begin with a GU and end with an AG. Presumably these short sequences assist in guiding the spliceosome.

Visual Evidence

The upper image is an electron micrograph of a mRNA-DNA hybrid molecule formed by mixing the messenger RNA (mRNA) from a clone of antibody-secreting cells with single-stranded DNA from the same kind of cells. The bar represents the length of 1000 bases.

The lower diagram is an interpretation of the micrograph. The solid line represents the DNA; the dotted line the mRNA. The loops (I_A, I_B, etc.) represent the introns that separate the exons encoding the domains of an antibody heavy chain:

- V = variable region
- E_1 = first constant region (C_H1) domain
- E_H = hinge region
- E_2 and E_3 = the nucleotides encoding the two C-terminal domains (C_H2 and C_H3).

The unhybridized portion of the mRNA is its poly(A) tail.

Alternative Splicing

The processing of pre-mRNA for many proteins proceeds along various paths in different cells or under different conditions. For example, early in the differentiation of a B cell (a lymphocyte that synthesizes an antibody) the cell first uses an exon that encodes a transmembrane domain that causes the molecule to be retained at the cell surface. Later, the B cell switches to using a different exon whose domain enables the protein to be secreted from the cell as a circulating antibody molecule.

Alternative splicing provides a mechanism for producing a wide variety of proteins from a small number of genes. While we humans may turn out to have only some 20 thousand genes, we probably make at least 10 times that number of different proteins.

It is now estimated that 92–94% of our genes produce pre-mRNAs that are alternatively-spliced. There is evidence that the pattern of alternative splicing differs consistently in different tissues and so must be regulated. But whether all the products are functional or that many are simply the outcome of an error-prone process remains to be seen.

Alternative splicing not only provides different proteins from a single gene but also different 3' UTRs and 5' UTRs. Although not translated into protein, these untranslated regions contain signals that, for example, dictate where in the cell that protein will accumulate. Two examples:

- The 3' UTR of the *bicoid* gene in Drosophila directs the mRNA to the anterior of the embryo;
- the same region in the *VegT* gene of Xenopus directs its mRNA to the vegetal pole of the embryo.

One of the most dramatic examples of alternative splicing is the *Dscam* gene in Drosophila. This single gene contains some 116 exons of which 17 are retained in the final mRNA. Some exons are always included; others are selected from an array. Theoretically this system is able to produce 38,016 different proteins. And, in fact, over 18,000 different ones have been found in Drosophila hemolymph.

These Dscam proteins are used to establish a unique identity for each neuron. It works like this. Each developing neuron draws upon the pool of thousands of possible different mRNAs to synthesize a dozen or so of them. Which ones are selected appears to be simply a matter of chance, but because of the size of the pool, each neuron will most likely end up with a unique set of a dozen or so Dscam proteins.

As each developing neuron in the central nervous system sprouts dendrites and an axon, these express its unique collection of Dscam proteins. If the extensions of a single neuron should meet each other in the tangled web that is the hallmark of nervous tissue, they are repelled. In this way, thousands of different neurons can coexist in intimate contact without the danger of nonfunctional contacts between the various extensions of the same neuron.

We also have a *DSCAM* gene which is located on chromosome 21. When present in 3 copies, it is probably responsible for some of the features of Down syndrome (and accounts for its full name: Down Syndrome Cell Adhesion Molecule). Perhaps the incredible diversity of synaptic junctions in the mammalian c.n.s. ($\sim 10^{14}$) is mediated by alternative splicing of a limited number of gene transcripts. So, whether a particular segment of RNA will be retained as an exon or excised as an intron can vary under different circumstances, such as;

- what type of cell the gene is in;
- what stage of differentiation that cell is passing through;
- what extracellular signals that cell is receiving.

Clearly the switching to an alternate splicing pathway must be closely regulated.

Trans-splicing

Most genes are transcribed and their transcripts processed as described above. RNA polymerase travels down a single strand of a singe gene locus to form pre-mRNA that is processed (including removal of introns) to form the mature mRNA. But there are exceptions. A number of cases have been found where two different precursor transcripts have been spliced together to form the final RNA molecule. The phenomenon is called *trans*-splicing. Examples: synthesis of a single RNA molecule by splicing together transcripts from ;

- loci located far apart on the same chromosome;
- opposite strands of the same gene locus;
- the two alleles of the gene on their separate (homologous) chromosomes.

The biological importance of these *trans*-spliced transcripts is still unknown for most of them.

Why Split Genes?

Perhaps during evolution, eukaryotic genes have been assembled from smaller, primitive genes - today's exons. Some proteins, like the antibodies mentioned in the previous section, are organized in a set of separate sections or domains each with a special function to perform in the complete molecule. Each domain is encoded by a separate exon. Having the different functional parts of the antibody molecule encoded by separate exons makes it possible to use these units in different combinations. Thus a set of exons in the genome may be the genetic equivalent of the various modular pieces in a box of "Lego" for children to assemble in whatever forms they wish. But the boundaries of other exons do not seem to correspond to domain boundaries of the protein. So perhaps the major benefit of split genes is simply the opportunity they provide for making many different proteins from a single gene through alternative splicing.

Summary

Gene expression occurs in two steps:

- transcription of the information encoded in DNA into a molecule of RNA (described here) and
- translation of the information encoded in the nucleotides of mRNA into a defined sequence of amino acids in a protein (discussed in Gene Translation: RNA → Protein).

Messenger RNA

Messenger RNA (mRNA) is a molecule of RNA encoding a chemical "blueprint" for a protein product. mRNA is transcribed from a DNA template, and carries coding information to the sites of protein synthesis: the ribosomes. Here, the nucleic acid polymer is translated into a polymer of amino acids: a protein. In mRNA as in DNA, genetic information is encoded in the sequence of nucleotides arranged into codons consisting of three bases each. Each codon encodes for a specific amino acid, except the stop codons that terminate protein synthesis. This process requires two other types of RNA: transfer RNA (tRNA) mediates recognition of the codon and provides the corresponding amino acid, while ribosomal RNA (rRNA) is the central component of the ribosome's protein manufacturing machinery.

Synthesis, Processing, and Function

The brief existence of an mRNA molecule begins with transcription and ultimately ends in degradation. During its life, an mRNA molecule may also be processed, edited, and transported prior to translation. Eukaryotic mRNA molecules often require extensive processing and transport, while prokaryotic molecules do not.

Transcription

During transcription, RNA polymerase makes a copy of a gene from the DNA to mRNA as needed. This process is similar in eukaryotes and prokaryotes. One notable difference, however, is that prokaryotic

RNA polymerase associates with mRNA processing enzymes during transcription so that processing can proceed quickly after the start of transcription. The short-lived, unprocessed or partially processed, product is termed *pre-mRNA*; once completely processed, it is termed *mature mRNA*.

Eukaryotic Pre-mRNA Processing

Processing of mRNA differs greatly among eukaryotes, bacteria and archea. Non-eukaryotic mRNA is essentially mature upon transcription and requires no processing, except in rare cases. Eukaryotic pre-mRNA, however, requires extensive processing.

5' Cap Addition

A *5' cap* (also termed an RNA cap, an RNA 7-methylguanosine cap or an RNA m^7G cap) is a modified guanine nucleotide that has been added to the "front" or 5' end of a eukaryotic messenger RNA shortly after the start of transcription. The 5' cap consists of a terminal 7-methylguanosine residue which is linked through a 5'-5'-triphosphate bond to the first transcribed nucleotide.

Its presence is critical for recognition by the ribosome and protection from RNases. Cap addition is coupled to transcription, and occurs co-transcriptionally, such that each influences the other. Shortly after the start of transcription, the 5' end of the mRNA being synthesized is bound by a cap-synthesizing complex associated with RNA polymerase. This enzymatic complex catalyzes the chemical reactions that are required for mRNA capping. Synthesis proceeds as a multi-step biochemical reaction.

Splicing

Splicing is the process by which pre-mRNA is modified to remove certain stretches of non-coding sequences called introns; the stretches that remain include protein-coding sequences and are called exons. Sometimes pre-mRNA messages may be spliced in several different ways, allowing a single gene to encode multiple proteins. This process is called alternative splicing. Splicing is usually performed by an RNA-protein complex called the spliceosome, but some RNA molecules are also capable of catalyzing their own splicing.

Ribozyme

A ribozyme (from ribonucleic acid enzyme, also called RNA enzyme or catalytic RNA) is an RNA molecule possessing a well defined tertiary structure that enables it to catalyze a chemical reaction. Many natural ribozymes catalyze either the hydrolysis of one of their own phosphodiester bonds (self-cleaving ribozymes), or the hydrolysis of bonds in other RNAs, but they have also been found to catalyze the aminotransferase activity of the ribosome. Investigators studying the origin of life have produced ribozymes in the laboratory that are capable of catalyzing their own synthesis under very specific conditions, such as an RNA polymerase ribozyme. Mutagenesis and selection has been performed resulting in isolation of improved variants of the "Round-18" polymerase ribozyme from 2001. "B6.61" is able to add up to 20 nucleotides to a primer template in 24 hours, until it decomposes by hydrolysis of its phosphodiester bonds.

Some ribozymes may play an important role as therapeutic agents, as enzymes which tailor defined RNA sequences, as biosensors, and for applications in functional genomics and gene discovery.

Discovery

Before the discovery of ribozymes, enzymes, which are defined as catalytic proteins, were the only known biological catalysts. In 1967, Carl Woese, Francis Crick, and Leslie Orgel were the first to suggest that RNA could act as a catalyst. This idea was based upon the discovery that RNA can form complex secondary structures. The first ribozymes were discovered in the 1980s by Thomas R. Cech, who was studying RNA splicing in the ciliated protozoan *Tetrahymena thermophila* and Sidney Altman, who was working on the bacterial RNase P complex.

These ribozymes were found in the intron of an RNA transcript, which removed itself from the transcript, as well as in the RNA component of the RNase P complex, which is involved in the maturation of pre-tRNAs. In 1989, Thomas R. Cech and Sidney Altman won the Nobel Prize in chemistry for their "discovery of catalytic properties of RNA." The term *ribozyme* was first introduced by Kelly Kruger *et al.* in 1982 in a paper published in *Cell.*

It had been a firmly established belief in biology that catalysis was reserved for proteins. In retrospect, catalytic RNA makes a lot of sense. This is based on the old question regarding the origin of life: Which comes first, enzymes that do the work of the cell or nucleic acids that carry the information required to produce the enzymes? Nucleic acids as catalysts circumvents this problem.

In the 1970s Thomas Cech, at the University of Colorado at Boulder, was studying the excision of introns in a ribosomal RNA gene in *Tetrahymena thermophila.* While trying to purify the enzyme responsible for splicing reaction, he found that intron could be spliced out in the absence of any added cell extract. As much as they tried, Cech and his colleagues could not identify any protein associated with the splicing reaction. After much work, Cech proposed that the intron sequence portion of the RNA could break and reform phosphodiester bonds. At about the same time, Sidney Altman, who is a Professor at Yale University, was studying the way tRNA molecules are processed in the cell when he and his colleagues isolated an enzyme called RNase-P, which is responsible for conversion of a precursor tRNA into the active tRNA.

Much to their surprise, they found that RNase-P contained RNA in addition to protein and that RNA was an essential component of the active enzyme. This was such a foreign idea that they had difficulty publishing their findings. The following year, Altman demonstrated that RNA can act as a catalyst by showing that the RNase-P RNA subunit could catalyze the cleavage of precursor tRNA into active tRNA in the absence of any protein component.

Since Cech's and Altman's discovery, other investigators have discovered other examples of self-cleaving RNA or catalytic RNA molecules. Many ribozymes have either a hairpin – or hammerhead – shaped active center and a unique secondary structure that allows them to cleave other RNA molecules at specific sequences. It is now possible to make ribozymes that will specifically cleave any RNA molecule. These RNA catalysts may have pharmaceutical applications. For example, a ribozyme has been designed to cleave the RNA of HIV. If such a ribozyme was made by a cell, all incoming virus particles would have their RNA genome cleaved by the ribozyme, which would prevent infection.

Activity

Although most ribozymes are quite rare in the cell, their roles are sometimes essential to life. For example, the functional part of the ribosome, the molecular machine that translates RNA into proteins, is fundamentally a ribozyme, composed of RNA tertiary structural motifs that are often coordinated to metal ions such as Mg^{2+} as cofactors. There is no requirement for divalent cations in a five-nucleotide RNA that can catalyze *trans*-phenylalanation of a four-nucleotide substrate which has three base complementary sequence with the catalyst. The catalyst and substrate were devised by truncation of the C3 ribozyme.

RNA can also act as a hereditary molecule, which encouraged Walter Gilbert to propose that in the distant past, the cell used RNA as both the genetic material and the structural and catalytic molecule, rather than dividing these functions between DNA and protein as they are today. This hypothesis became known as the "RNA world hypothesis" of the origin of life.

If ribozymes were the first molecular machines used by early life, then today's remaining ribozymes — such as the ribosome machinery — could be considered living fossils of a life based primarily on nucleic acids. A recent test-tube study of prion folding suggests that an RNA may catalyze the pathological protein conformation in the manner of a chaperone enzyme.

Known Ribozymes

Naturally occurring ribozymes include:

- Peptidyl transferase 23S rRNA
- RNase P
- Group I and Group II introns
- GIR1 branching ribozyme
- Leadzyme - Although initially created *in vitro*, natural examples have been found
- Hairpin ribozyme
- Hammerhead ribozyme
- HDV ribozyme
- Mammalian CPEB3 ribozyme
- VS ribozyme
- *glmS* ribozyme
- CoTC ribozyme.

Artificial Ribozymes

Since the discovery of ribozymes that exist in living organisms, there has been interest in the study of new synthetic ribozymes made in the laboratory. For example, artificially-produced self-cleaving RNAs that have good enzymatic activity have been produced.

Tang and Breaker isolated self-cleaving RNAs by in vitro selection of RNAs originating from random-sequence RNAs. Some of the synthetic ribozymes that were produced had novel structures, while some were similar to the naturally occurring hammerhead ribozyme. The techniques used to discover artificial ribozymes involve Darwinian evolution. This approach takes advantage of RNA's dual nature as both a catalyst and an informational polymer, making it easy for an investigator to produce vast populations of RNA catalysts using polymerase enzymes. The ribozymes are mutated by reverse transcribing them with reverse transcriptase into various cDNA and amplified with mutagenic PCR. The selection parameters in these experiments often differ. One approach for selecting a ligase ribozyme involves using biotin tags, which are covalently linked to the substrate. If a molecule possesses the desired ligase activity, a streptavidin matrix can be used to recover the active molecules.

Lincoln and Joyce developed an RNA enzyme system capable of self replication in about an hour. By utilizing molecular competition (*in vitro* evolution) of a candidate enzyme mixture, a pair of RNA enzymes emerged, in which each synthesizes the other from synthetic oligonucleotides, with no protein present.

Applications

A type of synthetic ribozyme directed against HIV RNA called gene shears has been developed and has entered clinical testing for HIV infection.

Editing

In some instances, an mRNA will be edited, changing the nucleotide composition of that mRNA. An example in humans is the apolipoprotein B mRNA, which is edited in some tissues, but not others. The editing creates an early stop codon, which upon translation, produces a shorter protein.

Polyadenylation

Polyadenylation is the addition of a poly(A) tail to an RNA molecule. The poly(A) tail consists of multiple adenosine monophosphates; in other words, it is a stretch of RNA which only has adenine bases. In eukaryotes, polyadenylation is part of the process that produces mature messenger RNA (mRNA) for translation. It therefore forms part of the larger process of gene expression.

The process of polyadenylation begins as the transcription of a gene finishes. The 3'-most segment of the newly-made RNA is first cleaved off by a set of proteins; these proteins then synthesize the poly(A) tail at the RNA's 3' end. In some genes these proteins may add a poly(A) tail at any one of several possible sites, polyadenylation can therefore produce more than one transcript from a single gene, similar to alternative splicing.

The poly(A) tail is important for the nuclear export, translation and stability of mRNA. The tail is shortened over time and when it is short enough, the mRNA is enzymatically degraded. However, in a few cell types, mRNAs with short poly(A) tails are stored for later activation by re-polyadenylation in the cytosol.

In contrast, when polyadenylation occurs in bacteria, it promotes RNA degradation. This is also sometimes the case for eukaryotic non-coding RNAs. The wide distribution of polyadenylation among living organisms indicates that this process evolved early in the history of life on Earth.

Primer on RNA

RNAs are a type of large biological molecules, whose individual building blocks are called nucleotides. The name *poly(A) tail* (for polyadenylic acid tail) reflects the way RNA nucleotides are abbreviated, with a letter for the base the nucleotide contains (A for adenine, C for cytosine, G for guanine and U for uracil). RNAs are produced (*transcribed*) from a DNA template. By convention, RNA sequences are written in a 5' to 3' direction. The 5' end is the part of the RNA molecule that is transcribed first, and the 3' end is transcribed last. The 3' end is also where the poly(A) tail is found on polyadenylated RNAs. Messenger RNA (mRNA) is RNA that has a coding region that acts as a template for protein synthesis (*translation*). The rest of the mRNA, the *untranslated regions*, tune how active the mRNA is. There are also many RNAs that are not translated, called non-coding RNAs. Like the untranslated regions, many of these non-coding RNAs have regulatory roles.

Nuclear Polyadenylation

Function

In nuclear polyadenylation, a poly(A) tail is added to an RNA at the end of transcription. On mRNAs, the poly(A) tail protects the mRNA molecule from enzymatic degradation in the cytoplasm and aids in transcription termination, export of the mRNA from the nucleus, and translation. Almost all eukaryotic mRNAs are polyadenylated, with the exception of animal replication-dependent histone mRNAs. These are the only mRNAs in eukaryotes that lack a poly(A) tail, ending instead in a stem-loop structure followed by a purine-rich sequence, termed histone downstream element, that directs where the RNA is cut so that the 3' end of the histone mRNA is formed.

Many eukaryotic non-coding RNAs are always polyadenylated at the end of transcription. There are small RNAs where the poly(A) tail is only seen in intermediary forms and not in the mature RNA as the ends are removed during processing, notably microRNAs. But for many long noncoding RNAs – a seemingly large group of regulatory RNAs that for example includes the RNA Xist which mediates X chromosome inactivation – a poly(A) tail is part of the mature RNA.

Mechanism

The polyadenylation machinery in the nucleus of eukaryotes works on products of RNA polymerase II, such as precursor mRNA. Here, a multi-protein complex cleaves the 3'-most part of a newly produced RNA and polyadenylates the end produced by this cleavage. The cleavage is catalysed by the enzyme CPSF and occurs 10–30 nucleotides downstream of its binding site.

This site is often the sequence AAUAAA on the RNA, but variants of it exist that bind more weakly to CPSF. Two other proteins add specificity to the binding to an RNA: CstF and CFI. CstF binds to a GU-rich region further downstream of CPSF's site. CFI recognises a third site on the RNA (a set of UGUAA sequences in mammals) and can recruit CPSF even if the AAUAAA sequence is missing. The polyadenylation signal – the sequence motif recognised by the RNA cleavage complex – varies between groups of eukaryotes. Most human polyadenylation sites contain the AAUAAA sequence, but this sequence is less common in plants and fungi.

The RNA is cleaved right after transcription, as CstF also binds to RNA polymerase II. Cleavage also involves the protein CFII, though it is unknown how. The cleavage site associated with a polyadenylation signal can vary up to some 50 nucleotides. When the RNA is cleaved, polyadenylation starts, catalysed by polyadenylate polymerase.

Polyadenylate polymerase builds the poly(A) tail by adding adenosine monophosphate units from adenosine triphosphate to the RNA, cleaving off pyrophosphate. Another protein, PAB2, binds to the new, short poly(A) tail and increases the affinity of polyadenylate polymerase for the RNA. When the poly(A) tail is approximately 250 nucleotides long the enzyme can no longer bind to CPSF and polyadenylation stops, thus determining the length of the poly(A) tail. CPSF is in contact with RNA polymerase II, allowing it to tell the polymerase to terminate transcription. The polyadenylation machinery is also physically linked to the spliceosome, a complex that removes introns from RNAs.

Downstream Effects

The poly(A) tail acts as the binding site for poly(A)-binding protein. Poly(A)-binding protein promotes export from the nucleus and translation, and inhibits degradation. This protein binds to the poly(A) tail prior to mRNA export from the nucleus and in yeast also recruits poly(A) nuclease, an enzyme that shortens the poly(A) tail and allows the export of the mRNA. Poly(A)-binding protein is exported to the cytoplasm with the RNA. mRNAs that are not exported are degraded by the exosome. Poly(A)-binding protein also can bind to, and thus recruit, several proteins that affect translation, one of these is initiation factor-4G which in turn recruits the 40S ribosomal subunit. However, a poly(A) tail is not required for the translation of all mRNAs.

Deadenylation

In eukaryotic somatic cells, the poly(A) tail of most mRNAs in the cytoplasm gradually get shorter, and mRNAs with shorter poly(A) tail are translated less and degraded sooner. However, it can take many hours before an mRNA is degraded. This deadenylation and degradation process can be accelerated by microRNAs complementary to the 3' untranslated region of an mRNA. In immature egg cells, mRNAs with shortened poly(A) tails are not degraded, but are instead stored without being translated.

They are then activated by cytoplasmic polyadenylation after fertilisation, during egg activation. In animals, poly(A) ribonuclease (PARN) can bind to the 5' cap and remove nucleotides from the poly(A) tail. The level of access to the 5' cap and poly(A) tail is important in controlling how soon the mRNA is degraded. PARN deadenylates less if the RNA is bound by the initiation factors 4E (at the 5' cap) and 4G (at the poly(A) tail), so this is why translation reduces deadenylation. The rate of deadenylation may also be regulated by RNA-binding proteins. Once the poly(A) tail is removed, the decapping complex removes the 5' cap, leading to a degradation of the RNA. Several other enzymes that seem to be involved in deadenylation have been identified in yeast.

Alternative Polyadenylation

Many protein-coding genes have more than one polyadenylation site, so a gene can code for several mRNAs that differ in their 3' end. Since alternative polyadenylation changes the length of the 3' untranslated region, it can change which binding sites for microRNAs the 3' untranslated region contains. MicroRNAs tend to repress translation and promote degradation of the mRNAs they bind to, although there are examples of microRNAs that stabilise transcripts. Alternative polyadenylation can also shorten the coding region, thus making the mRNA code for a different protein, but this is much less common than just shortening the 3' untranslated region. The choice of poly(A) site can be influenced by extracellular stimuli and depends on the expression of the proteins that take part in polyadenylation.

For example, the expression of CstF-64, a subunit of cleavage stimulatory factor (CstF), increases in macrophages in response to lipopolysaccharides (a group of bacterial compounds that trigger an immune response). This results in the selection of weak poly(A) sites and thus shorter transcripts.

This removes regulatory elements in the 3' untranslated regions of mRNAs for defense-related products like lysozyme and TNF-α. These mRNAs then have longer half-lives and produce more of these proteins. RNA-binding proteins other than those in the polyadenylation machinery can also affect whether a polyadenyation site is used,as can DNA methylation near the polyadenylation signal.

Cytoplasmic Polyadenylation

There is polyadenylation in the cytosol of some animal cell types, namely in the germ line, during early embryogenesis and in post-synaptic sites of nerve cells. This lengthens the poly(A) tail of an mRNA with a shortened poly(A) tail, so that the mRNA will be translated. These shortened poly(A) tails are often less than 20 nucleotides, and are lengthened to around 80–150 nucleotides.

In the early mouse embryo, cytoplasmic polyadenylation of maternal RNAs from the egg cell allows the cell to survive and grow even though transcription does not start until the middle of the 2-cell stage (4-cell stage in human). In the brain, cytoplasmic polyadenylation is active during learning and could play a role in long-term potentiation, which is the strengthening of the signal transmission from a nerve cell to another in response to nerve impulses and is important for learning and memory formation.

Cytoplasmic polyadenylation requires the RNA-binding proteins CPSF and CPEB, and can involve other RNA-binding proteins like Pumilio. Depending on the cell type, the polymerase can be the same type of polyadenylate polymerase (PAP) that is used in the nuclear process, or the cytoplasmic polymerase GLD-2.

Tagging for Degradation in Eukaryotes

For many non-coding RNAs, including tRNA, rRNA, snRNA and snoRNA, polyadenylation is a way of marking the RNA for degradation, in at least yeast.

This polyadenylation is done in the nucleus by the TRAMP complex, which adds a tail that is around 40 nucleotides long to the 3' end. The RNA is then degraded by the exosome. Poly(A) tails have also been found on human rRNA fragments, both the form of homopolymeric (A only) and heterpolymeric (mostly A) tails.

In Prokaryotes and Organelles

In many bacteria, both mRNAs and non-coding RNAs can be polyadenylated. This poly(A) tail promotes degradation by the degradosome, which contains two RNA-degrading enzymes: polynucleotide phosphorylase and RNase E. Polynucleotide phosphorylase binds to the 3' end of RNAs and the 3' extension provided by the poly(A) tail allows it to bind to the RNAs whose secondary structure would otherwise block the 3' end. Successive rounds of polyadenylation and degradation of the 3' end by polynucleotide phosphorylase allows the degradosome to overcome these secondary structures.

The poly(A) tail can also recruit RNases that cut the RNA in two. These bacterial poly(A) tails are about 30 nucleotides long. In as different groups as animals and trypanosomes, the mitochondria contain both stabilising and destabilising poly(A) tails.

Destabilising polyadenylation targets both mRNA and noncoding RNAs. The poly(A) tails are 43 nucleotides long on average. The stabilising ones start at the stop codon, and without them the stop codon (UAA) is not complete as the genome only encodes the U or UA part. Plant mitochondria only have destabilising polyadenylation, and yeast mitochondria have no polyadenylation at all.

While many bacteria and mitochondria have polyadenylate polymerases, they also have another type of polyadenylation, performed by polynucleotide phosphorylase itself.

This enzyme is found in bacteria, mitochondria, plastids and as a constituent of the archeal exosome (in those archaea that have an exosome). It can synthesise a 3' extension where the vast majority of the bases are adenines. Like in bacteria, polyadenylation by polynucleotide phosphorylase promotes degradation of the RNA in plastids and likely also archaea.

Evolution

Although polyadenylation is seen in almost all organisms, it is not universal. However, the wide distribution of this modification and the fact that it is present in organisms from all three domains of life implies that the last universal common ancestor of all living organisms probably had some form of polyadenylation system. A few organisms do not polyadenylate mRNA, which implies that they have lost their polyadenylation machineries during evolution. Although no examples of eukaryotes that lack polyadenylation are known, mRNAs from the bacterium *Mycoplasma gallisepticum* and the salt-tolerant archaean *Haloferax volcanii* lack this modification.

The most ancient polyadenylating enzyme is polynucleotide phosphorylase. This enzyme is part of both the bacterial degradosome and archaeal exosome, two closely related complexes that recycle RNA into nucleotides. This enzyme degrades RNA by attacking the bond between the 3'-most nucleotides with a phosphate, breaking off a diphosphate nucleotide. This reaction is reversible, and so the enzyme can also extend RNA with more nucleotides.

The heteropolymeric tail added by polynucleotide phosphorylase is very rich in adenine. The choice of adenine is most likely the result of higher ADP concentrations than other nucleotides as a result of using ATP as an energy currency, making it more likely to be incorporated in this tail in early lifeforms. It has been suggested that the involvement of adenine-rich tails in RNA degradation prompted the later evolution of polyadenylate polymerases (the enzymes that produce poly(A) tails with no other nucleotides in them).

Polyadenylate polymerases are not as ancient. They have separately evolved in both bacteria and eukaryotes from CCA-adding enzyme, which is the enzyme that completes the 3' ends of tRNAs. Its catalytic domain is homologous to that of other polymerases. Likely, the horizontal transfer of bacterial CCA-adding enzyme to eukaryotes allowed the archaeal-like CCA-adding enzyme to switch function to a poly(A) polymerase. Some lineages, like archaea and cyanobacteria, never evolved a polyadenylate polymerase.

History

Polyadenylation was first identified in 1960 as an enzymatic activity in extracts made from cell nuclei that could polymerise ATP, but not ADP, into polyadenine. Although identified in many types of cells, this activity had no known function until 1971, when poly(A) sequences were found in mRNAs. Initially, the only function of these sequences was thought to be protection of the 3' end of the RNA from nucleases, but later the specific roles of polyadenylation in nuclear export and translation were identified. The polymerases responsible for polyadenylation were first purified and characterized in the 1960s and 1970s, but the large number of accessory proteins that control this process were only discovered in the early 1990s. Polyadenylation is the covalent linkage of a polyadenylyl moiety to a messenger RNA molecule. In eukaryotic organisms, most messenger RNA (mRNA) molecules are polyadenylated at the 3' end.

The poly(A) tail and the protein bound to it aid in protecting mRNA from degradation by exonucleases. Polyadenylation is also important for transcription termination, export of the mRNA from the nucleus, and translation. mRNA can also be polyadenylated in prokaryotic organisms, where poly(A) tails act to facilitate, rather than impede, exonucleolytic degradation. Polyadenylation occurs during and immediately after transcription of DNA into RNA. After transcription has been terminated, the mRNA chain is cleaved through the action of an endonuclease complex associated with RNA polymerase. After the mRNA has been cleaved, around 250 adenosine residues are added to the free 3' end at the cleavage site. This reaction is catalyzed by polyadenylate polymerase. Just as in alternative splicing, there can be more than one polyadenylation variant of a mRNA.

Transport

Another difference between eukaryotes and prokaryotes is mRNA transport. Because eukaryotic transcription and translation is compartmentally separated, eukaryotic mRNAs must be exported from the nucleus to the cytoplasm. Mature mRNAs are recognized by their processed modifications and then exported through the nuclear pore. In neurons mRNA must be transported from the soma to the dendrites where local translation occurs in response to external stimuli. Many messages are marked with so-called "zip codes" which targets their transport to a specific location.

Translation

Because prokaryotic mRNA does not need to be processed or transported, translation by the ribosome can begin immediately after the end of transcription. Therefore, it can be said that prokaryotic translation is *coupled* to transcription and occurs *co-transcriptionally*.

Eukaryotic mRNA that has been processed and transported to the cytoplasm (i.e. mature mRNA) can then be translated by the ribosome. Translation may occur at ribosomes free-floating in the cytoplasm, or directed to the endoplasmic reticulum by the signal recognition particle. Therefore, unlike in prokaryotes, eukaryotic translation *is not* directly coupled to transcription.

Structure

5' cap

The 5' cap is a specially altered nucleotide on the 5' end of precursor messenger RNA and some other primary RNA transcripts as found in eukaryotes. The process of 5' capping is vital to creating mature messenger RNA, which is then able to undergo translation. Capping ensures the messenger RNA's stability while it undergoes translation in the process of protein synthesis, and is a highly regulated process that occurs in the cell nucleus.

5' Cap Structure

The 5' cap is found on the 5' end of an mRNA molecule and consists of a guanine nucleotide connected to the mRNA via an unusual 5' to 5' triphosphate linkage. This guanosine is methylated on the 7 position directly after capping *in vivo* by a methyl transferase. It is referred to as a 7-methylguanosine cap, abbreviated m^7G.

Further modifications include the possible methylation of the 2' hydroxy-groups of the first 2 ribose sugars of the 5' end of the mRNA. The methylation of both 2' hydroxy-groups is shown on the diagram. The 5' cap looks like the 3' end of an RNA molecule (the 5' carbon of the cap ribose is bonded, and the 3' unbonded). This provides significant resistance to 5' exonucleases.

Capping Process

The starting point is the unaltered 5' end of an RNA molecule. This features a final nucleotide followed by three phosphate groups attached to the 5' carbon.

1. One of the terminal phosphate groups is removed (by RNA terminal phosphatase), leaving two terminal phosphates.
2. GTP is added to the terminal phosphates (by a guanylyl transferase), losing two phosphate groups (from the GTP) in the process. This results in the 5' to 5' triphosphate linkage.
3. The 7-nitrogen of guanine is methylated (by a methyl transferase).
4. Other methyltransferases are optionally used to carry out methylation of 5' proximal nucleotides.

5' Capping Targeting

The Capping Enzyme Complex (CEC) required for capping is found bound to the RNA polymerase II before transcription starts. As soon as the 5' end of the new transcript emerges the enzymes transfer to it and begin the capping process (this is a similar kind of mechanism to ensure capping as for polyadenylation). The enzymes for capping can only bind to RNA polymerase II ensuring specificity to only these transcripts, which are almost entirely mRNA.

5' Cap Function

The 5' cap has 4 main functions:

1. Regulation of nuclear export.
2. Prevention of degradation by exonucleases.
3. Promotion of translation.
4. Promotion of 5' proximal intron excision.

Nuclear export of RNA is regulated by the Cap binding complex (CBC), which binds exclusively to capped RNA. The CBC is then recognized by the nuclear pore complex and exported. Once in the cytoplasm after the pioneer round of translation, the CBC is replaced by the translation factors eIF-4E and eIF-4G. This complex is then recognized by other translation initiation machinery including the ribosome.

Cap prevents 5' degradation in two ways. First, degradation of the mRNA by 5' exonucleases is prevented (as mentioned above) by functionally looking like a 3' end. Second, the CBC complex and the eIF-4E/eIF-4G block the access of decapping enzymes to the cap. This increases the half-life of the mRNA, essential in eukaryotes as the export process takes significant time.

Decapping of an mRNA is catalyzed by the decapping complex made up of at least Dcp1 and Dcp2, which must compete with eIF-4E to bind the cap.

Thus the 5' cap is a marker of an actively translating mRNA and is used by cells to regulate mRNA half-lives in response to new stimuli. Undesirable mRNAs are sent to P-bodies for temporary storage or decapping, the details of which are still being resolved.

The mechanism of 5' proximal intron excision promotion is not well understood, but the 5' cap appears to loop around and interact with the spliceosome in the splicing process, promoting intron excision.

Messenger RNA Decapping

The process of messenger RNA decapping consists of hydrolysis of the 5' cap structure on the RNA exposing a 5' monophosphate. This 5' monophosphate is a substrate for the exonuclease Xrn1 and the message is quickly destroyed. There are many situations which may lead to the removal of the cap, some of which are discussed below.

Translation and Decay

Inside cells there is a balance between the processes of translation and mRNA decay. Messages which are being actively translated are bound by polysomes and the initiation factors eIF-4E and eIF-4G. This blocks access to the cap by the decapping enzyme DCP2 and protects the message.

In nutrient-starvation conditions or viral infection, translation may be compromised and decapping is stimulated. This balance is reflected in the size and abundance of the cytoplasmic structures known as P-bodies

Specific Decay Pathways

A number of specific decay pathways exist that recognize aberrant messages and promote their decapping. Nonsense mediated decay recognizes premature stop codons and promotes decapping as well as decay by the exosome. Certain classes of miRNA have also been shown to stimulate decapping.

Coding Regions

The coding region of a gene is that portion of a gene's DNA or RNA, composed of exons, that codes for protein. The region is bounded nearer the 5' end by a start codon and nearer the 3' end with a stop codon. The coding region in mRNA is bounded by the five prime untranslated region and the three prime untranslated region, which are also parts of the exons.

The coding region of an organism is sum total of the organism's genome that is composed of gene coding regions. Coding regions are composed of codons, which are decoded and translated (in eukaryotes usually into one and in prokaryotes usually into several) proteins by the ribosome.

Coding regions begin with the start codon and end with a stop codon. Generally, the start codon is an AUG triplet and the stop codon is UAA, UAG, or UGA.

The coding regions tend to be stabilised by internal base pairs, this impedes degradation. In addition to being protein-coding, portions of coding regions may serve as regulatory sequences in the pre-mRNA as exonic splicing enhancers or exonic splicing silencers.

Open Reading Frame

In molecular genetics, an open reading frame (ORF) is a DNA sequence that does not contain a stop codon in a given reading frame.

Significance

One common use of open reading frames are as one piece of evidence to assist in gene prediction. Long ORFs are often used, along with other evidence, to initially identify candidate protein coding regions in a DNA sequence. The presence of an ORF does not necessarily mean that the region is ever translated.

For example in a randomly generated DNA sequence with an equal percentage of each nucleotide, a stop-codon would be expected once every 21 codons. A simple gene prediction algorithm for prokaryotes might look for a start codon followed by an open reading frame that is long enough to encode a typical protein, where the codon usage of that region matches the frequency characteristic for the given organism's coding regions. By itself even a long open reading frame is not conclusive evidence for the presence of a gene.

Example

If a portion of a genome has been sequenced (e.g. 5'-ATCTAAAATGGGTGCC-3'), ORFs can be located by examining each of the three possible reading frames on each strand. In this sequence two out of three possible reading frames are entirely open, meaning that they do not contain a stop codon:

1. ...A TCT AAA ATG GGT GCC...
2. ...AT CTA AAA TGG GTG CC...
3. ...ATC TAA AAT GGG TGC C...

Possible stop codons in DNA are "TGA", "TAA" and "TAG". Thus, the last reading frame in this example contains a stop codon (*TAA*), unlike the first two.

Untranslated Regions

Untranslated regions (UTRs) are sections of the mRNA before the start codon and after the stop codon that are not translated, termed the five prime untranslated region (5' UTR) and three prime untranslated region (3' UTR), respectively. These regions are transcribed with the coding region and thus are exonic as they are present in the mature mRNA. Several roles in gene expression have been attributed to the untranslated regions, including mRNA stability, mRNA localization, and translational efficiency.

The ability of a UTR to perform these functions depends on the sequence of the UTR and can differ between mRNAs. The stability of mRNAs may be controlled by the 5' UTR and/or 3' UTR due to varying affinity for RNA degrading enzymes called ribonucleases and for ancillary proteins that can promote or inhibit RNA degradation.

Translational efficiency, including sometimes the complete inhibition of translation, can be controlled by UTRs. Proteins that bind to either the 3' or 5' UTR may affect translation by influencing the ribosome's ability to bind to the mRNA. MicroRNAs bound to the 3' UTR also may affect translational efficiency or mRNA stability.

Cytoplasmic localization of mRNA is thought to be a function of the 3' UTR. Proteins that are needed in a particular region of the cell can actually be translated there; in such a case, the 3' UTR may contain sequences that allow the transcript to be localized to this region for translation.

Some of the elements contained in untranslated regions form a characteristic secondary structure when transcribed into RNA. These structural mRNA elements are involved in regulating the mRNA. Some, such as the SECIS element, are targets for proteins to bind. One class of mRNA element, the riboswitches, directly bind small molecules, changing their fold to modify levels of transcription or translation. In these cases, the mRNA regulates itself.

5' UTR

The five prime untranslated region (5' UTR), also known as the leader sequence, is a particular section of messenger RNA (mRNA) and the DNA that codes for it.

At times the leader sequence can contain elements for controlling gene expression by way of attenuation. The leader sequence will be translated and the availability of the encoded amino acids will affect the transcription of the rest of the mRNA.

It starts at the +1 position (where transcription begins) and ends one nucleotide before the start codon (usually AUG) of the coding region.

It usually contains a ribosome binding site (RBS), in bacteria also known as the Shine Dalgarno sequence (AGGAGGU). The 5' UTR may be a hundred or more nucleotides long, and the 3' UTR may be even longer (up to several kilobases in length).

An mRNA molecule codes for a protein through translation. The mRNA also contains regions that are not translated: in eukaryotes this includes the 5' untranslated region, 3' untranslated region, 5' cap and poly-A tail. In prokaryotic mRNA the 5' UTR is normally short. Some viruses and cellular genes have unusual long structured 5' UTRs which may have roles in gene expression.

Several regulatory sequences may be found in the 5' UTR:

- Binding sites for proteins, that may affect the mRNA's stability or translation, for example iron responsive elements, which occur in the 5' UTRs (and 3' UTRs) of a small number of eukaryotic mRNAs that regulate gene expression in response to iron.
- Regulatory elements that do not depend on proteins, such as riboswitches.
- Sequences that promote the initiation of translation.

3' UTR

The three prime untranslated region (3' UTR) is a particular section of messenger RNA (mRNA). It follows the coding region.

An mRNA molecule codes for a protein through translation. The mRNA also contains regions that are not translated. In eukaryotes these regions are the cap, 5' untranslated region, 3' untranslated region, and polyA tail.

In prokaryotes mRNA structures have some differences as do histone mRNAs. However, both have 3' UTRs.

Several regulatory sequences are found in the 3' UTR:

- A polyadenylation signal, usually *AAUAAA*, or a slight variant. This marks the site of cleavage of the transcript approximately 30 base pairs past the signal, followed by the addition of several hundred adenine residues (poly-A tail).
- Binding sites for proteins, that may affect the mRNA's stability or location in the cell, like SECIS elements (which direct the ribosome to translate the codon UGA as selenocysteines rather than as a stop codon), or AU-rich elements (AREs), stretches consisting of mainly adenine and uracil nucleotides (which can either stabilize or destabilize the mRNA depending on the protein bound to it).
- Binding sites for miRNAs, a type of RNAi.

Poly(A) Tail

The 3' poly(A) tail is a long sequence of adenine nucleotides (often several hundred) added to the 3' end of the pre-mRNA. This tail promotes export from the nucleus and translation, and protects the mRNA from degradation.

Monocistronic versus Polycistronic mRNA

An mRNA molecule is said to be monocistronic when it contains the genetic information to translate only a single protein. This is the case for most of the eukaryotic mRNAs. On the other hand, polycistronic mRNA carries the information of several genes, which are translated into several proteins.

These proteins usually have a related function and are grouped and regulated together in an operon. Most of the mRNA found in bacteria and archea are polycistronic. Dicistronic or bicistronic is the term used to describe an mRNA that encodes only two proteins.

mRNA Circularization

In eukaryotes it is thought that mRNA molecules form circular structures due to an interaction between the cap binding complex and poly(A)-binding protein. Circularization is thought to promote recycling of ribosomes on the same message leading to efficient translation.

Degradation

Different mRNAs within the same cell have distinct lifetimes (stabilities). In bacterial cells, individual mRNAs can survive from seconds to more than an hour; in mammalian cells, mRNA lifetimes range from several minutes to days.

The greater the stability of an mRNA, the more protein may be produced from that mRNA. The limited lifetime of mRNA enables a cell to alter protein synthesis rapidly in response to its changing needs. There are many mechanisms that lead to the destruction of a mRNA, some of which are described below.

Prokaryotic mRNA Degradation

In prokaryotes the lifetime of mRNA is generally much shorter than in eukaryotes. Prokaryotes degrade messages by using a combination of ribonucleases, including endonucleases, 3' exonucleases, and 5' exonucleases. In some instances, small RNA molecules (sRNA) tens to hundreds of nucleotides long can stimulate the degradation of specific mRNAs by base pairing with complementary sequences and facilitating ribonuclease cleavage. It was recently shown that bacteria also have a sort of 5' cap consisting of a triphosphate on the 5' end. Removal of two of the phosphates leaves a 5' monophosphate, causing the message to be destroyed by the endonuclease RNase E.

Eukaryotic mRNA Turnover

Inside eukaryotic cells there is a balance between the processes of translation and mRNA decay. Messages that are being actively translated are bound by ribosomes, the eukaryotic initiation factors eIF-4E and eIF-4G, and poly(A)-binding protein. eIF-4E and eIF-4G block the decapping enzyme (DCP2), and poly(A)-binding protein blocks the exosome complex, protecting the ends of the message. The balance between translation and decay is reflected in the size and abundance of cytoplasmic structures known as P-bodies

The poly(A) tail of the mRNA is shortened by specialized exonucleases that are targeted to specific messenger RNAs by a combination of cis-regulatory sequences on the RNA and trans-acting RNA-binding proteins. Poly(A) tail removal is thought to disrupt the circular structure of the message and destabilize the cap binding complex. The message is then subject to degradation by either the exosome complex or the decapping complex. In this way, translationally inactive messages can be destroyed quickly, while active messages remain intact.

AU-rich Element Decay

The presence of AU-rich elements in some mammalian mRNAs tends to destabilize those transcripts through the action of cellular proteins that bind these sequences and stimulate poly(A) tail removal. Loss of the poly(A) tail is thought to promote mRNA degradation by facilitating attack by both the exosome complex and the decapping complex. Rapid mRNA degradation via AU-rich elements is a critical mechanism for preventing the overproduction of potent cytokines such as tumour necrosis factor (TNF) and granulocyte-macrophage colony stimulating factor (GM-CSF). AU-rich elements also regulate the biosynthesis of proto-oncogenic transcription factors like c-Jun and c-Fos.

Nonsense Mediated Decay

Nonsense-mediated decay (NMD) is a cellular mechanism of mRNA surveillance that functions to detect nonsense mutations and prevent the expression of truncated or erroneous proteins. Following transcription, precursor mRNA undergoes an assemblage of ribonucleoprotein (RNP) components followed by regulatory pre-mRNA processing. Large average intron size in eukaryotic cells greatly increases the probability that aberrant mRNA splicing will result in the presence of a nonsense (stop) codon (UAA, UAG, UGA) somewhere within the open reading frame. NMD is triggered by exon junction complexes (EJCs) (components of the assembled RNP) that are deposited during pre-mRNA processing. While EJCs located in an open reading frame upstream of exon-exon junctions may facilitate ribosomal recruitment prior to displacement by a "pioneer" round of translation, EJCs located downstream of a nonsense codon are not displaced because the ribosome is released from the transcript before reaching it. These remaining EJCs function as tags for recruitment of UPF1 following the mRNA's transport out of the nucleus and into the cytosol where the RNA is degraded, for example by the exosome complex. NMD is not only a mechanism for degrading aberrant mRNA, however, as there are numerous examples of normal transcripts whose expression is regulated by this process including the plasticity protein Arc/Arg3.1.

Eukaryotic messages are subject to surveillance by nonsense mediated decay (NMD), which checks for the presence of premature stop codons (nonsense codons) in the message. These can arise via incomplete splicing, V(D)J recombination in the adaptive immune system, mutations in DNA, transcription errors, leaky scanning by the ribosome causing a frame shift, and other causes. Detection of a premature stop codon triggers mRNA degradation by 5' decapping, 3' poly(A) tail removal, or endonucleolytic cleavage.

Chapter 10

Physiology: Plant Growth and Development

The three major plant functions that are the basics for plant growth and development are photosynthesis, respiration, and transpiration.

Photosynthesis

One of the major differences between plants and animals on earth is the ability of plants to internally manufacture their own food. To produce food for itself a plant requires energy from sunlight, carbon dioxide from the air and water from the soil.

If any of these ingredients is lacking, photosynthesis, or food production, will stop. If any factor is removed for a long period of time, the plant will die. Photosynthesis literally means "to put together with light."

$$\text{Carbon dioxide} + \text{Water} \xrightarrow{\text{(Light)}} \text{Sugar} + \text{Oxygen}$$

$$6CO_2 6H_2O \xrightarrow{\text{(Light)}} C_6H_{12}O_6 + 6O_2$$

Any green plant tissue is capable of photosynthesis. Chloroplasts in these cells contain the green pigment called chlorophyll which traps the light energy. However, leaves are generally the site of most food production due to their special structure.

The internal tissue (mesophyll) contains cells with abundant chloroplasts in an arrangement that allows easy movement of water and air. The protective upper and lower epidermis (skin) layers of the leaf include many stomata that are openings in the leaf formed by two specialized guard cells on either side. Guard cells regulate movement of the gases, (i.e. CO_2 into and O_2 and H_2O out of the leaf), involved in photosynthesis. The lower epidermis of the leaf normally contains the largest percentage of stomata.

Light Reaction

Photosynthesis is the process of turning the energy of sunlight into chemical energy from the raw products of CO_2 and H_2O. This process is necessary to sustain nearly all forms of life. Photosynthesis is divided in to two separate reactions known as the light and dark reactions. They take place when light is present but the dark reaction does not require light. The whole process is begun by light reacting with pigments in the leaf causing the splitting of water molecules. This is called photolysis or the *Hill Reaction* which is not completely understood. Three products are produced in this reaction. Electrons from the hydrogen molecules and remaining H^+ ions are used to form two separate energy storage molecules. The air we breath is from the remaining oxygen portion of H_2O. The carbon dioxide molecules are transformed into sugars during the dark reaction using the energy that was formed during the light reaction.

Dark Reaction

This part of the photosynthetic process is also called the Calvin Cycle. With one cycle of this reaction 3 carbon atoms are fixed or placed in a sugar molecule. This pathway is called C-3 photosynthesis. This is the way that most dicots or broadleaf plants make sugars during the dark reaction. C-3 photosynthesis has a disadvantage though. Oxygen competes with CO_2 for a binding site during the dark reaction. Sometimes sugars are not formed, but energy is still expended to complete the cycle. This is called photorespiration.

Another dark reaction pathway is called C-4 photosynthesis because 4 carbons are fixed or placed in a sugar molecule each time the cycle is completed. The dark reaction of C-4 photosynthesis occurs inside of specialized parts of leaf cells in the leaf called the bundle sheath, which exclude the presence of O_2. Because there is no oxygen present photorespiration does not occur. The C-4 photosynthetic pathway is what occurs in most monocots or grasses. This is a more efficient pathway and allows grasses to grow faster than broadleaf plants. Crassulacean acid metabolism or CAM photosynthesis is the dark reaction type found in many cactus, succulents, bromeliads, and orchids as well as a few other plants. CAM photosynthesis is similar to C-4 photosynthesis. However, CAM plants open their stomata only during the night to collect CO_2, when air temperatures are cooler, thus conserving water because of reduced transpiration. The CO_2 is converted into malic acid and then converted back to CO_2 during the day when light is present, thus producing sugars, while the stomata are closed and greatly reducing water loss.

Plants convert the energy from light into simple sugars, such as glucose. This food may be converted back to water and carbon dioxide, releasing the stored energy through a process called respiration. This energy is required for growth in nearly all organisms. Simple sugars are also converted to other sugars and starches (carbohydrates) which may be transported to the stems and roots for use or storage, or they may be used as building blocks for more complex structures, e.g. oils, pigments, proteins, cell walls, etc.

Photosynthesis is dependent on the availability of light. Generally speaking, as sunlight increases in intensity photosynthesis increases. This results in greater food production. Many garden crops, such as tomatoes, respond best to maximum sunlight. Tomato production is cut drastically as light intensities drop. Only two or three varieties of "greenhouse" tomatoes will produce any fruit when sunlight is minimal in fall and spring.

Water plays an important role in photosynthesis in several ways. First, it maintains a plant's turgor or the firmness or fullness of plant tissue. Turgor pressure in a cell can be compared to air in an inflated balloon. Water pressure or turgor is needed in plant cells to maintain shape and ensure cell growth. Second, water is split into hydrogen and oxygen by the energy of the sun that has been absorbed by the chlorophyll in the plant leaves. The oxygen is released into the atmosphere and the hydrogen is used in manufacturing carbohydrates. Third, water dissolves minerals from the soil and transports them up from the roots and throughout the plant, where they serve as raw materials in the growth of new plant tissues.

The soil surrounding a plant should be moist, not too wet or too dry. Water is pulled through the plant by evaporation of water through the leaves (transpiration). Photosynthesis also requires carbon dioxide (CO_2) which enters the plant through the stomata. Carbon and oxygen are used in the manufacture of carbohydrates. Carbon dioxide in the air is 350 parts per million (ppm) or 0.035% at sea level and is plentiful enough so that it is not a limiting factor in plant growth. However, since carbon dioxide is consumed in making sugars and is not replenished by plants at a rapid rate, a tightly closed greenhouse in midwinter may not let in enough outside air to maintain an adequate carbon dioxide level. Under these conditions, improved crops of roses, carnations, tomatoes and certain other crops can be produced if the carbon dioxide level is raised with CO_2, generators or, in small greenhouses, with dry ice or a natural gas flame. Although not a direct component in photosynthesis, temperature is an important factor. Photosynthesis occurs at its highest rate in the temperature range of 65° to 85°F (18° to 27°C) and decreases when temperatures are above or below this range.

Respiration

Carbohydrates made during photosynthesis are of value to the plant when they are converted into energy. This energy is used in the process of building new tissues. The chemical process by which sugars and starches produced by photosynthesis are converted into energy is called respiration. It is similar to the burning of wood or coal to produce heat or energy. This process in cells is shown most simply as:

$$C_6H_{12}O_6 + 6O_2 \rightarrow 6CO_2 + 6H_2O + \text{Energy}$$

$$\text{Sugar} + \text{Oxygen} \rightarrow \text{Carbon dioxide} + \text{Water}$$

This equation is precisely the opposite of that used to illustrate photosynthesis, although more is involved than just reversing the reaction. However, it is appropriate to relate photosynthesis to a building process, while respiration is a breaking-down process.

Respiration:

1. Uses food for plant energy.
2. Releases energy.
3. Occurs in all cells.
4. Uses oxygen.
5. Produces water.
6. Produces carbon dioxide.
7. Occurs in darkness as well as light.

If oxygen is limited or not present then anaerobic respiration or metabolism occurs. The by products of this reaction are ethyl alcohol or lactic acid and CO_2. This process is also know as fermentation or the Pasteur effect, (Louis Pasteur was the first to describe the effect), which is used to manufacture brewing and dairy products. It also occurs in muscle tissue when they are over exerted. The muscle burning we feel doing exercises is the accumulated lactic acid that forms in our tissue because of limited oxygen. Plant tissues undergo the same process, for example waterlogged soils limit the oxygen available to roots and may cause them to rot because of fermentation.

Photosynthesis:

1. Produces food.
2. Stores energy.
3. Occurs in cells containing chloroplasts.
4. Releases oxygen.
5. Uses water.
6. Uses carbon dioxide.
7. Occurs in sunlight.

By now, it should be clear that respiration is the reverse of photosynthesis. Unlike photosynthesis, respiration occurs at night as well as during the day. Respiration occurs in all life forms and in all cells. The release of accumulated carbon dioxide and the uptake of oxygen occurs at the cell level. In animals, blood carries both oxygen and carbon dioxide to and from the atmosphere by means of the lungs, gills, spiracles etc. In plants there is simple diffusion into the open spaces within the leaf and exchange occurs through the stomata.

Transpiration

Transpiration is the process by which a plant loses water, primarily through leaf stomata. Transpiration is a necessary process that involves the use of about 90% of the water that enters the plant through the roots. The other 10% of the water is used in chemical reactions and in plant tissues. Transpiration is necessary for mineral transport from the soil to the plant for the cooling of the plant through evaporation, to move sugars and plant chemicals, and for the maintenance of turgor pressure. The amount of water lost from the plant depends on several environmental factors such as temperature, humidity and wind or air movement. An increase in temperature or air movement decreases relative humidity and causes the guard cells in the leaf to shrink, opening the stomata and increasing the rate of transpiration.

Environmental Factors that affect Plant Growth

Plant growth and distribution are limited by the environment. If any one environmental factor is less than ideal it will become a limiting factor in plant growth. Limiting factors are also responsible for the geography of plant distribution. For example, only plants adapted to limited amounts of water can live in deserts. Most plant problems are caused by environmental stress, either directly or indirectly. Therefore, it is important to understand the environmental aspects that affect plant growth. These factors are light, temperature, water (humidity), and nutrition.

Light

Light has three principal characteristics that affect plant growth: quantity, quality, and duration.

Light quantity refers to the intensity or concentration of sunlight and varies with the season of the year. The maximum is present in the summer and the minimum in winter. The more sunlight a plant receives (up to a point), the better capacity it has to produce plant food through photosynthesis. As the sunlight quantity decreases the photosynthetic process decreases. Light quantity can be decreased in a garden or greenhouse by using shade-cloth or shading paint above the plants. It can be increased by surrounding plants with white or reflective material or supplemental lights.

Light quality refers to the colour or wavelength reaching the plant surface. Sunlight can be broken up by a prism into respective colours of red, orange, yellow, green, blue, indigo, and violet. On a rainy day, raindrops act as tiny prisms and break the sunlight into these colours producing a rainbow. Red and blue light have the greatest effect on plant growth. Green light is least effective to plants as most plants reflect green light and absorb very little. It is this reflected light that makes them appear green. Blue light is primarily responsible for vegetative growth or leaf growth. Red light when combined with blue light, encourages flowering in plants. Fluorescent or cool-white light is high in the blue range of light quality and is used to encourage leafy growth. These lights are excellent for starting seedlings. Incandescent light is high in the red or orange range but generally produces too much heat to be a valuable light source. Fluorescent "grow" lights have a mixture of red and blue colours that attempts to imitate sunlight as closely as possible. They are costly and generally not of any greater value than regular fluorescent lights.

Light duration or photoperiod refers to the amount of time that a plant is exposed to sunlight. When the concept of photoperiod was first recognized it was thought that the length of periods of light triggered flowering. The various categories of response were named according to the light length (i.e., short-day and long-day). It was then discovered that it is not the length of the light period but the length of uninterrupted dark periods that is critical to floral development. The ability of many plants to flower is controlled by photoperiod.

Plants can be classified into three categories, depending upon their flowering response to the duration of darkness. These are short-day, long-day, or day-neutral plants. Short-day, (long nights) plants form their flowers only when the day length is less than about 12 hours in duration. Short-day plants include many spring and fall flowering plants such as chrysanthemum and poinsettia. Long-day, (short nights) plants form flowers only when day lengths exceed 12 hours. They include almost all of the summer-flowering plants, as well as many vegetables including beet, radish, lettuce, spinach, and potato.

Day-neutral plants form flowers regardless of day length. Some plants do not really fit into any category but may be responsive to combinations of day lengths. The petunia will flower regardless of day length, but flowers earlier and more profusely under long daylight. Since chrysanthemums flower under the short-day conditions of spring or fall the method for manipulating the plant into experiencing short days is very simple. If long days are predominant, a black plastic sheet is drawn over the chrysanthemum for 12 hours daily to block out light until flower buds are initiated. To bring a long-day plant into flower when sunlight is not present longer than 12 hours artificial light is added until flower buds are initiated.

Temperature

Temperature affects the productivity and growth of a plant depending upon whether the plant variety is a warm-season or cool-season crop. If temperatures are high and day length is long, cool-season crops such as broccoli and spinach will bolt rather than produce the desired flower. Temperatures that are too low or high for a warm-season crop will prevent fruit set. Temperatures that are too high for warm-season crops such as pepper or tomato can cause pollen to become inviable and not pollinate flowers. Adverse temperatures also cause stunted growth and poor quality. For example, the bitterness in lettuce is caused by high temperatures.

Sometimes temperatures are used in connection with day length to manipulate the flowering of plants. Chrysanthemums will flower for a longer period of time if daylight temperatures are 59°F (15°C). The Christmas cactus forms flowers as a result of short days and low temperatures. Temperatures alone also influence flowering. Daffodils are forced to flower by putting the bulbs in cold storage in October at 35° to 40°F (2° to 4°C). The cold temperatures allow the bulb to mature. The bulbs are transferred to the greenhouse in midwinter where growth begins. The flowers are then ready for cutting in 3 to 4 weeks.

Thermoperiod refers to daily temperature change. Plants produce maximum growth when exposed to a day temperature that is about 10 to 15° F. (5.5 to 8°C) higher than the night temperature. This allows the plant to photosynthesize and respire during an optimum daytime temperature and to curtail the rate of respiration during a cooler night.

High temperatures cause increased respiration sometimes above the rate of photosynthesis. This means that the products of photosynthesis are being used more rapidly than they are being produced. For growth to occur photosynthesis must be greater than respiration.

Low temperatures can result in poor growth. Photosynthesis slows at low temperatures. Since photosynthesis is slowed, growth is slowed and this results in lower yields. Not all plants grow best in the same temperature range. For example, snapdragons grow best when nighttime temperatures are 55°F (12°C); the poinsettia prefers 62°F (17°C). Florist cyclamen does well under very cool conditions while many bedding plants prefer a higher temperature. Recently it has been found that roses can tolerate much lower nighttime temperatures than was previously believed. This has meant a conservation in energy for greenhouse growers.

However, in some cases a certain number of days of low temperatures are needed by plants to grow properly. This is true of crops growing in cold regions of the country. Peaches are a prime example; most varieties require 700 to 1,000 hours below 45°F (7°C) and above 32°F (0°C) before they break their rest period and begin flowering and growth. If this cold requirement is not met then small, misshapen leaves and fruit will result. Many times fruit will not set. In low desert areas where these temperatures are not experienced low chill peach trees should be planted. Lilies need 6 weeks at 33°F (1°C or below) before they will bloom.

Plants can be classified as either hardy or non-hardy depending upon their ability to withstand cold temperatures. Winter injury can occur to non-hardy plants if temperatures are too low or if unseasonably low temperatures occur late in the spring or early in the fall. Winter injury may also occur because of desiccation (drying out).

Plant roots need moist soil during the winter. When the soil is frozen the movement of water into the plant is severely restricted. On a windy winter day broad-leaved evergreens can become water-deficient in a few minutes, turning the leaves or needles brown. Wide variations in winter temperatures can cause premature bud break in some plants and consequent freezing damage. Late spring frost damage can ruin entire crops. If temperatures drop too low during the winter, entire trees of some species are killed by the freezing of plant cells and tissue.

Phytopathology

Plants like all organisms can become diseased. They grow old, they are attacked by parasitic fungi, insects, bacterial and viral infections. In addition to these factors, they also can be affected by the alleopathic effects of other plants, disruption of root space and changes in the water table.

Most plant diseases have either cures or controls. Controls being the preferred option. Diseases can be prevented before they occur, by numerous methods. These measures may be time consuming and expensive. Therefore, for ornamental plants, people generally prefer to attempt a cure after diseases become manifest. Not all diseases are so easy to treat. Consider *chestnut blight.*

Mountain Pine Beetle

The mountain pine beetle, Rocky Mountain beetle, or Black Hills beetle (*Dendroctonus ponderosae*), is a small beetle, less than 5 mm long, that bores into the bark of pine trees. There it lays its eggs, the hatching larvae feed in the bark, i.e. the phloem, of pine trees.

It over-winters as a larva in the bark. In the northern States and Canada, the beetle usually has but one complete generation per year. The beetle infests mostly the lodgepole, limber and ponderosa pines. Western white pine, Scots pine, and other *Pinus* species can be affected also. Mostly these beetle infestations occur in the far west, and it is stands of lodgepole and ponderosa pine that are most often decimated. It has been found that the beetle is capable of feeding off some eastern pines – if given the chance. It can potentially live in jack pine.

Outbreaks tend to be localised, but large in overall extent. It is not usually the mountain pine beetle itself that kills a pine tree. It is rather a series of 'blue stain' fungi that do this injury (eg. *Ophiostoma clavigerum* & *Ophiostoma montium*). These fungi start in the phloem, but the mycelia can eventually spread into the sapwood.

If they grow too prolifically they can block the tree's sap-flow- thereby killing the tree. As is often the case, these fungi live symbiotically with the beetle. The beetle larva eats the fungal mycelium, but the adult spreads the fungal spores. The adult beetle spreads the spores of the fungus when its seeks out a new host tree.

Mountain pine beetle is native to western North America. However, pine die-off has reached 'pandemic' (panphytic?) proportions in the last decade. Vast swaths of pine forest have been killed so far by the beetle-fungus combo.

It seems that global warming may have decreased the winter mortality of the beetle. With more beetles surviving winter, woodpeckers and other natural contrals cannot always keep up. Outbreaks may also be exacerbated by the shorter dormant season for which the pines are perhaps somewhat maladapted. The blue stain fungus mostly kills trees stressed by some other cause. The beetle most afflicts old trees, dought stressed trees and/or over-crowded stands.

Systemic insecticides can control the beetle. Controlled burning of moribund trees does work. Sometimes burning is the only workable option. Diligence is needed to stop the beetle and its fungi from colonizing the jack pines in the boreal forest to the east.

Chestnut Blight

American Sweet Chestnuts *(Castanea dentata)* are trees in the beech-oak family. Sweet chestnuts were once common in the *Carolinian* deciduous forests of eastern North America. In some forests *associations* they were the dominant species, often a quarter or more of the trees per stand were sweet chestnuts. Sweet chestnuts were most often associates with black oak, white oak, shagbark hickory, white ash and black cherry.

There are other species of chestnut. The European *Castanea sativa*, and the Asian *Castanea mollissima*, are similar and produce edible seeds. These species are, however, less damaged by chestnut blight than their American relative.

In 1904 a fungal disease (*Cryphonectria parasitica*) of Asian origin was introduced, by accident, to New York State. The fungus growing in the thin sapwood of the sweet chestnuts encountered insufficient resistance from the trees' defences. By 1925 this fungus infestation spread to almost the whole of the sweet chestnut's native range. Eventually, almost every sweet chestnut stand was infected. Isolated trees, and some coppice sprouts, are still producing seeds. The coppice sprouts are not good at re-establishing roots. And the old stumps are being slowly rotted away by the chestnut-tongue fungus (*Fistulina hepatica*). However, a few adult trees still exist. This author has seen blight-free chestnuts near the coast of Lake Erie. Perhaps they are there protected from blight spores by the direction of the wind. All hope is not lost.

Cryphonectria parasitica is a canker fungus. It causes lesions on branches or stems. On other tree species, such as oaks and hickories, the fungus seems to be purely saprophytic, meaning that it feeds on wood that is already dead. To the genus *Castanea*, however, the fungus is truly parasitic, it manages to kill living wood, and then it digests the now-dead tissue. It is an *ascomycete* fungus and produces ascospores, sexually produced spores. Ascospores are released from a small disc-like 'perithecia' bodies that develop on the bark. These ascospores are generally wind-dispersed. This fungus also produces conidia, which are asexual spores. These conidia are present in a sticky orange tendrils that ooze from the conidiospore producing bodies called 'pycnidia'. Conidia are most often spread by insects, birds and other mobile vectors that come in contact with the sticky ooze. Like many ascomycete species, the blight also has deuteromycete forms. Some strains of the chestnut blight do not seem to produce ascospores, only conidiospores. These asexual anamorphs are known as *Endothiella parasitica.*

Control

Several scientists are still trying to breed blight resistant strains of chestnut. For, unfortunately, there appears to be no effective post-infection treatment for chestnut blight. So far, there has been little sign of progress.

A similar fate has fallen on the butternut, the white elm, and the ash trees. All of trees are threatened by fungal diseases which have been spread by human agency. Likewise, several European and Asian plants have been infected by fungi of American origin. If there is any object lesson in this, it is to be careful with introducing organisms to places where they are not native. Often this mixus cannot be avoided.

Dutch Elm Disease

Dutch elm disease, or 'elm wilt', is more famous, now, than chestnut blight, because the epidemic was nearer in time to our era, and because the effects were less drastic overall.

Dutch elm disease is caused by either of two species of fungus, *Ophiostoma ulmi* and *Ophiostoma novo-ulmi.* These fungi are members of the *Plectomycetes* series, and are distinguished by having *cleistothecia,* which is to say ascospores enclosed in a 'sac'. The *Ophiostoma* fungi are very yeast like in form. Toxins produced by the fungus are carried up conductive vessels, where they cause leaf wilt.

The fungal mycelia also block the conducting elements, causing the wood to die. These fungi are spread by bark beetles, usually the native *Hylurgopinus rufipes*, or the European *Scolytus multistriatus* beetle. Both beetles, as larvae, chew galleries just below the bark in the outermost annual ring. Each gallery has a main chamber from which radiates a series of blind tunnels.

The eggs were laid in the main chamber, and the individual larvae each chew out one of the side tunnels. The *Ophiostoma* fungi use the beetles as vectors, the spores growing on the inside of the tunnels rub off on the larvae, and more importantly they brush onto the adults that emerge from the pupae.

Apparently spores can also be eaten by the beetles, and survive passage through their digestive tracts. The pupated adults then fly off, bearing spores which may infect the elm on which the beetles lay their eggs. The fungi are introduced to a new elm when the beetle lays its eggs in the bark of a new tree. The fungi then attack the sapwood of that new elm.

Since elms only have a few actively conducting annual rings, the fungi in this thin sapwood have a greater impact than they would if the sapwood layers were thicker. The fungal mycelia do not remain confined to the vicinity of the larval galleries. The fungus spread throughout the sapwood, including the conductive wood of the roots. Sometimes the mycelia can cross from the roots of one elm to another. Elms often have their roots fused with neighbouring elms.

Elms infested with the 'Dutch elm disease' may die in the same year as they are infected. Often they die slowly over the span of a few years. Older trees are more likely to be infected than young ones. Most elms, except for a few Asian species, are sensitive to the Dutch elm disease fungus. However, the largest native elm, the American elm or white elm (*Ulmus americana*), is very susceptible to Dutch elm disease. Dutch elm disease originated in Asia, spread to Europe and then to North America. Probably the disease arrived on a shipment of lumber. It made its first appearance in Ohio in 1930. By the 1950s thousands of white elms were dying throughout North America. By 1970 most truly large elms were dead giant hulks with slabs of bark peeling off. The disease seems to have flared up again in the 1990s, as some of the small surviving elms grew to medium size, and then caught the disease.

White elm did not go into a sharp decline, rather it was that old well formed trees became increasingly uncommon. Many white elm saplings live long enough to produce a seed crop before they are killed by the fungus. Some of the Asian elms are less susceptible, including Siberian elm which is a popular city tree. There are several wild species of elm also, each with varying ability to withstand the fungus. There are, in short, plenty of wild and domestic elms on which the fungus can perpetuate itself.

The fungus is not 'burning itself out'. Nevertheless, the white elm could be taking the long slow route to extinction. Several people, including Bernd Heinrich of the University of Vermont, have suggested that white elm could survive Dutch elm disease. It has been noted that some elms develop seed earlier than others. This is due to ordinary genetic variation. Since Dutch elm disease afflicts older trees more often than young ones, there is therefore natural selection for white elms that flower at a younger age. Perhaps, white elm is evolving from a tree into a large shrub.

Control

There have been attempts to find resistant strains of white elm. There are systemic fungicide injections that can control the fungus. These must be repeated every year or so, and they are fairly expensive. New methods aim, instead, to stimulate and enhance the natural defences of the elm trees. The process is like 'inoculation'. Although, the 'immune system' of plants is much simpler than the immune system of animals. Plants do build-up chemical defences to irritants. Thus, one can 'trick' the plant into mounting a defensive response, with a chemical stimulus. Since the defences build up in the absence of an actual fungal attack, the elm becomes pre-prepared for a real fungus infestation. If there are many elms to treat, systemic fungicides and inoculants may be an impractical solution to the problem. In practice, this means that white elms usually die, and must be removed by a qualified arborist. The cost of removal may exceed $1600.00 cdn, if the elm is large.

Butternut Canker

In 1991 a canker disease was noticed in butternuts (*Juglans cinerea*). Within a few years it was to be found decimating butternuts, a tree that was never common to begin with. It is believed that the fungus is an introduced species. Though, this is not certain.

The wilt is caused by fungus called *Sirococcus clavigignenti-juglandacearum*. Like most canker fungi it causes wounds that heal slowly, if at all. It also causes a telltale slime-flux or gummosis. And like too many fungal diseases it cannot be easily eradicated once a tree is infected. Luckily, some strains and hybrids are fairly resistant to the fungus.

The situation has gotten serious enough that in Ontario butternuts are on the Endangered Species List. Though, many states in the USA have not taken this step – yet. If you suspect butternut canker in your neighbourhood, call the nearest Urban Forestry Service, or a professional arborist. They can help with diagnosis and suggestions for further action.

Pine Diplodia

'Diplodia' is now the common name for the *Shaeropsis sapinea* fungus. 'Diplodia' is the former generic name of the *Shaeropsis sapinea* fungus. The *Diplodia* genus still exists, but is occupied by other closely related fungi. *Shaeropsis sapinea* is a *Coelomycetes* fungus. It is best known for its infestations of firs and pines, in particular the Austrian pine (*Pinus nigra*). Austrian pine is a European species that is now a common urban tree in North America.

Symptoms of diplodia infestations begin with a yellowing of the new tip leaves (needles) of a pine. By the summer and autumn of the year of yellowing, leaves fall in great masses. (Diplodia is sometimes considered a 'needle cast' fungus.)

If a branch becomes denuded of leaves, the whole branch dies. Often lower branches show these symptoms first, and then they spread to the upper crown. Although, this sequence does not always follow. In Austrian pine infestations of diplodia can defoliate the tree in the second or third year after the onset of symptoms. Before this, several branches in the mid crown may be dead or dying.

Unfortunately, Austrian pines have in the past often been planted together in large numbers. Residential and industrial sites may have virtual forests of such pines. It is not uncommon for several trees to develop diplodia in one year and spread the disease to others the next. Consequently, with a span of two years, dead and dying pines are scattered all over the property. Tree removal professionals may like the results of diplodia, but property managers generally do not.

Control

Although, in theory it is possible to stay the effects of the fungus with injection of fungicides, in practice this is seldom done. For even if successful, the fungus has probably spread to other trees, and almost certainly large portions of the crown have been disfigured already. Old Austrian pines tend to be more often afflicted than the young. Indeed, diplodia very often kills Austrian pines which are near the end of their life cycle.

Austrian Pine

Austrian pine, Corsican pine or black pine (*Pinus nigra*) is a sturdy pine. The leaves occur in bundles of two, they are dark, long, stiff and bristle tipped. The crown of a healthy Austrian pine is very thickly foliated. The cones are large and rounded. The pine can grow up to 30 metres tall. Austrian pine occurs in nature in the southern Apennine, and the Mediterranean highlands in Greece, Corsica and Sicily. Austrian pines are not long lived, nor are they hardy in most of North America.

American seasons are generally more extreme in temperature, and drought stress, than are the seasons in Europe. Quite often, in fact, drought stress, root compaction, and other physical stresses, precede the onset of diplodia infection. Thus, 'stress' is said to make the pines more susceptible to disease in general. This 'stress' explanation has become a cliché, some scepticism is called for. Watering and fertilising, although generally good for a tree, probably will not much improve the chance of survival for diplodia infected Austrian pine.

Austrian pines were once a favoured urban tree species, because they are cheap and fast growing. Their short life span, and susceptibility to diplodia make them a poor urban tree choice. They also have very large cones, which are a hassle for groundskeepers and lawn mowers. Cost considerations often tempt developers and property buyers to choose plant Austrian pines as ornamental trees for residential and industrial sites. My recommendation is: do not plant Austrian pine.

Witches' Brooms

Witches' brooms are clusters of stunted, closely packed distorted twigs growing on trees or shrubs. From a distance the tangled mass of a witch broom may look like a mistletoe, squirrel's nest or a bird's nest. Usually these deformities are confined to a small region on a plant. Witch broom deformities are instigated by several different agents, viral, phytoplasmal and fungal. In all cases the infectious agent apparently distorts the normal growth pattern of the plants. The growth tends to follow the same architectural programme as normal tissue, but the rates of growth, the ultimate size and the symmetry of the plant are disrupted.

Taphrina spp. are *Ascomycete* fungi that can cause witches' brooms on birch (*Betula spp.*), plum (*Prunus spp.*) and hornbeam (*Carpinus spp.*). Different *Taphrina* species afflict different host plants.

Those *Taphrina* species that do not cause witch broom distortions usually cause leaf, flower or fruit distortions. Some *Taphrina* species cause *leaf curl* diseases, leaves that are cup-shaped and not flat. Others cause twisted asymmetrical fruit, as manifest in the *pocket plum* disease.

Melampsorella caryophyllacearum is a 'rust' or *Uredinales* fungus. It can cause cankers or witch broom deformities on fir trees (*Abies spp.*). Like many rust fungi, the 'fir rust' has alternate hosts.

Fir rust can infect chickweed (*Cerastium*) and stitchwort (*Stellaria*). Like other rust fungi, different kinds of spores, sexual and asexual, occur on the different hosts. The *aeciospores* produced on the fir originate from the union of 'male' *spermatia* and 'female' *receptive hyphae.*

Phytoplasmas are another agent that can cause witches' brooms to develop on locust (*Robina spp.*) and apple (*Malus spp.*). Phytoplasmas are tiny monera similar to bacteria.

Control

There is little evidence that witches' brooms cause any serious harm to a tree. However, they are sometimes considered unsightly. If one insists on removing witches' brooms, pruning off the offending broom is an effective treatment. Winter is the best time for removing witches' brooms. It is recommended that witches' brooms be burnt after they are removed.

Bedeguar Galls

Sometimes around mid-summer one can find strange mossy growths on roses (*Rosa spp.*). These look like tiny deformed tendrils all bunched up in a mass of several centimetres width. Usually they are pallid green, with strands of red and yellow. These are 'bedeguar' galls.

Bedeguar galls are galls instigated by wasps (*Diplolepis rosae*). The tiny female wasps lay their eggs inside young buds. The growing buds form the deformed bedeguar growths. The larvae dwell within little chambers at the centres of these growths. The adults do not emerge from these galls until the following spring. Bedeguar galls, once established, can only be controlled by pruning them off. Luckily, they do not afflict trees in the rose family.

Viroids, Viruses and Phytoplasmas

Graft-Transmissible Agents

Plants are often infected with viruses and viroids. Many of these viruses have been specifically identified. For others, their existence has been surmised on the basis of virus-like symptoms. When no identified virus or bacterium has been verified, botanists often call the disease a '*graft-transmissible agent*'. The term refer to the fact that the agent can be transmitted by grafting, or contact between plants, just like a virus. It is suspected that some graft-transmissible agents are viroids.

Mosaic Viruses

A virus is an microorganism that exists as a complete cellular parasite. A virus utilises the cellular mechanisms of their host for their replication. They have either DNA or RNA based gene sequences that are 'read' by the host cell. The host cell is 'tricked' into replicating the virus' genome and its proteins. Viruses are capable of very limited self-activated responses, such as penetrating the host cell membrane. Viruses and graft-transmissible agents generally either cause deformed growth, or mottling of leaves. Deformities may include stunted growth, or misshapen roots, stems, leaves, flowers or fruit. Leaf mottling is an especially common symptom of virus infection. Each 'mottle' being a location where virus replication causes either tissue necrosis or interruption in cell function. If the mottling is tight it forms a 'mosaic' pattern, hence the term *mosaic virus*.

There are 'viruses' with some of their own functional genes. These are the Mimiviridae which include the *Mimivirus*. These weird creatures are hundreds of times more massive than most 'normal' viruses. The mimiviruses are built like small bacteria. They are parasites mostly of protozoa and animals.

Viroids

Viroids are basically viruses that lack a protein coat. Many viroids have been identified, the genomes of several have even been mapped. Viroids consist only of strands of RNA. Viroids have the shortest and simplest known genetic sequences of any 'living' thing. One viroid, the *potato spindle tuber virus* (PSTV) that infects potatoes, has only 359 nucleotide bases in its genome! A viroid that infests oranges, *citrus exocortis* (CEV), has only 371 nucleotide bases. Like viruses, viroids are complete cellular parasites, they require living cells to replicate their genes for them.

Phytoplasmas-MLOs

Some graft-transmissible agents which were originally thought to be viroids or viruses have actually been identified as *phytoplasmas*. Phytoplasmas are basically very tiny *Gram Positive* bacteria in the

Class *Mollicutes*. They are similar to the *mycoplasmas* that cause a number of diseases in animals. Phytoplasmas were once commonly known as "mycoplasma like organisms" (MLOs).

Phytoplasmas lack cell walls, but they do have cell membranes and an active metabolism. Their nucleotide sequences have several thousand base pairs. They are obligate parasites in that they lack the ability to live without their hosts. Consequently, they have reduced amounts of DNA and proteins within them. Phytoplasmas are quite small, only a few hundred nanometres wide (70-1000 nm). This is just larger than the enigmatic nanobacteria.

Phytoplasmas are known to cause a number of diseases in plants. They live inside their host cells' cytoplasm. Many MLOs can be spread from plant to plant via sap-feeding insects. Symptoms of phytoplasmic disease often involve disturbances in the growth of plant tissues. For example, *apple chat* phytoplasma causes apples to cease development before they rippen properly. *Apple proliferation* phytoplasma causes 'witches brooms', a tangle of distorted twigs and leaves. *Apple rubbery wood* phytoplasma causes flexible rubbery shoots.

These are just a few examples that infest apples. Others phytoplasma diseases infect roses, strawberries, and various garden vegetables. Some very important forest diseases are MLOs. Elm yellows and ash yellows are caused by phytoplasmas. There are presently no effective cures for phytoplasma diseases. The spread of these bacters can be controlled by removing and burning infected twigs. Avoid any sort of contact between the infected plant tissues and healthy plants.

Black Knot

Black knot (*Apiosporina morbosa*) is a fungus that attacks the cherries, plums, apricots, almonds and other species in the genus *Prunus*. (An old name for the fungus was: *Dibotyron morbosum*). Infections start in wounds in the bark. Usually the visible symptoms are manifest the following year. Hard black rough growths swell up around the wound and expand into sausage shaped masses. The 'knots' can both grow and spread into uninfected twigs. In the spring olive-green spores form on the knots.

Control

If the fungal knots start on large stems, or spread to them, the damage may become difficult to rectify. Knot-cankers on trunks can kill the tree. The best method of control is to prune off the infected twigs several centimetres back from the knot. These cuttings mst be disposed of, ideally they should be burnt. At any rate, pruning tools must be sterilised before being used on other trees.

Damping-Off

Damping-off is a catch-all term for soil fungoids which cause butt-rot in seedlings. Oomycetes in the *Pythium* genus are a widespread cause of this rot.

The damping-off fungoid usually waits dormant in the soil in spore form, 'hatching' only when a suitable host plant becomes available.

Damping-off infection is visible as a blackening of bark near the root crown. If the rot is extensive enough, the seedling may die. Usually humid environments favour the damping-off infestation.

The disease is a problem mostly, but not exclusively, in greenhouses. In an urban forestry context, it is mostly a problem for the tree-nursery, not the homeowner.

Control: One can sterilise the greenhouse compost, with heat or fumigants, before seeding. Ultimately, soil should not be too vigorously sterilised, as beneficial mycorrhizae could be killed in the process. In general, damping-off can be minimised by good soil hygiene, and by not allowing the seeding milieu to become too damp.

Mildews & Moulds

'Mildew' is a catch-all term for organisms which cause mouldy rots. Mildews, as a whole, are somewhat more common in moist shaded areas than in dry areas with bright sunlight. Pathogenic mildews are not easily distinguished from pathogenic moulds (U.S. spelling: 'mold'). The term 'mildew', to a plant pathologist, signifies the parasitic downy and powdery mildews.

Biofilms

Various kinds of bacteria, slime mould amoebae, yeast fungi and even algae can form 'scums' on plant surfaces. The slime of these biofilms can be composed of a gelatinous matrix of water, polysaccharides and proteins.

The organisms excrete the gelatine to protect their little microbial community. Sometimes the slimes are composed of several species. Some of the species may be mutualistic, others may be predatory. Biofilms on plants are usually not pathogenic in themselves.

Generally, they are merely symptomatic of an extremely humid milieu. They can, however, be precursors to true mould infections.

Downy Mildew

Downy mildews are oomycete fungoids in the *Peronospraceae* family. The branched sporangia are large enough to appear, to the unaided-eye, as downy or cottony masses. These obligate parasites feed on a wide range of plant tissues. Most species are very host specific. They are not as likely to be a major landscape problem as are the powdery mildews.

Control: Oomycetes are not easily controlled by copper fungicides. (They are not fungi.) *Mancozeb* is somewhat effective in killing the fungoid.

Powdery Mildew

Powdery mildews are ascomycete fungi in the *Erysiphaceae* family. The masses of mycelium on plant tissues may look to the human eye like a fine white to grey powder. Generally the fruiting-bodies are smaller than those of the downy mildews. The fine mycelia feed on plant cells, weakening them in an overall general sense. Distortions of growth may occur if young growing tissues are infected.

Control: Sulphur based fungicides can inhibit the growth of powdery mildew. However, several applications during dry weather are usually necessary. Synthetic fungicides are more effective and require fewer treatments.

Fungi Imperfecti

The *Deuteromycota*, or *Fungi Imperfecti*, are those fungi which originally were not assignable to specific taxonomic groups. They could not be assigned because they had no distinguishing visible features that could identify them as either basidomycetes, ascomycetes or other fungal taxa. Increasingly, genetic studies have identified the taxonomic affinities of the imperfect fungi.

For example, the so-called '*Botrytis cinerea*' mould is actually a race of the ascomycete *Botryotina fuckeliana*. The grey mould form does not produce ascospores. The *Botryotina fuckeliana* form does produce such spores.

The *Monilia brunnea* fungus which lives symbiotically with ambrosia beetles was an imperfect fungi. Its genetic affiliation links it to the ascomycetes. Ascomycetes are somewhat more prone to being imperfect fungi than are the basidiomycetes. Those imperfect fungi which seem to lack even a conidial phase are called *mycelia sterilia* (i.e. sterile mould). Due to the vast numbers of imperfect fungi species, the reclassification process is still incomplete.

Sooty Mould (Mold)

Sooty moulds are an assorted set of saprobic fungi. They appear as dark mouldy patches on leaves, twigs or stems. These fungi are not directly disease agents. They feed off the faeces of aphids, and other insects with sugary excretions. The insects, not the fungi, are more likely to be the actual health problem.

Control: Sooty moulds are usually confined to regions of the plant with sap feeding insects on them. If sap-feeders are identified as the culprits, the insects need to be controlled, not the fungi.

Grey Mould-Gray Mold

Grey mould (*gray mold*) is a fungus that consist of mycelium strands growing on and within plant tissues. In moist conditions the fungus can begin to grow on leaves, twigs and fruits. Basically, it causes soft tissues to rot. 'Grey mould' is a general name. Many different fungal species have been labelled 'grey moulds'. The pseudo-generic name *Botrytis* has been given to many of the *Fungi Imperfecti* that cause grey mould.

Control: Sulphur based fungicides can kill grey mould, if applied in time. Increasing ventilation so as to dry the plants also helps check the spread of the fungus. Drying is not always practical for outdoor plants. The best advice is to trim off dead branch stubs, as the fungal attack often starts in dead tissues.

Bibliography

Ashraf, Syed Mahboob: *Handbook of Fruit and Vegetable Products*, Agrobios, Delhi, 2008.

Ausubel, F.M. : *Current Protocols in Molecular Biology,* New York: John Wiley and Sons, 1989.

Barbeau G.: *Tropical Fruits in Nicaragua*, Managua, Nicaragua Ministerio de Desarrollo Agropecuario, Agraria, 1990.

Brodwin, Paul : *Biotechnology and Culture: Bodies, Anxieties, Ethics*, Bloomington: Indiana University Press, 2001.

Chadha K. L. and Pareek O. P.: *Advances in Horticulture: Fruit Crops,* New Delhi, Malhotra Publishing House, 1993.

Currah L. and Proctor F. J.: *Onions in Tropical Regions*, Kent, Natural Resources Institute, 1990.

Dharamvir Hota: *Modern Biotechnology in Plant Breeding*, Gene Tech Books, Delhi, 2007.

Dudley, E.: *The Critical Villager: Beyond Community Participation*, London, Routledge, 1993.

Elliott, B N : *Biochemistry and Molecular Biology*, Oxford University Press, Delhi, 2005.

Fortuner, R.: *Nematode Identification and Expert System Technology,* New York, Plenum Press, 1988.

Fumento, Michael: *Bioevolution: How Biotechnology is Changing Our World*, San Francisco, Encounter Books, 2003.

Gangawane L.V. and Khilare V.C. : *Molecular Biology of Plant Pathogens*, Daya, Delhi, 2010.

Georges, S.: *The Debt Boomerang: How Third World Debt Harms Us All*, Boulder, Westview Press, 1992.

Hardy B.: *Biology and Agronomy of Forage Arachis*, Cali, International Centre for Tropical Agriculture, 1994.

Jeffers P.: *Evaluation of Four Onion Varieties in Montserrat*, Plymouth, CARDI, 1992.

Jha Timir Baran and Ghosh Biswajit : *Plant Tissue Culture : Basic and Applied*, Universities Press, Delhi, 2005.

Kuppuram, G & K. Kumudamani: *History of Science and Technology in India*, Delhi, Sundeep Prakashan, 1990.

Lorenzen, S.: *The Phylogenetic Systematics of Freeliving Nematodes*, London, The Ray Society, 1994.

Matsuoka H.: *Cultivation of Panicum Genetic Resources for Evaluation of Characteristics*, Tokyo, JICA, 1997.

Morris M. L. : *Maize Seed Industries in Developing Countries*, Colorado, Lynne Rienner, 1998.

Nobel, P. S.: *Physicochemical and Environmental Plant Physiology*, Academic Press, San Diego, 1999.

Nobel, P. S.: *Physicochemical and Environmental Plant Physiology*, Academic Press, San Diego, 1999.

Oldham P.: *Cost of Production of Major Tree Crops in Dominica*, Roseau, Ministry of Agriculture, 1991.

Panno, Joseph : *Stem Cell Research: Medical Applications and Ethical Controversy,* New York: Checkmark, 2006.

Politycka, B and C L Goswami: *Plant Physiology : Research Methods*, Scientific, Delhi, 2007.

Qystein V. Sjaastad: *Physiology of Domestic Animals*, International Book Distributing Co., Delhi, 2005.

Ragone D.: *Breadfruit: Artocarpus Altilis (Parkinson) Fosberg*, Rome, International Plant Genetic Resources Institute, 1997.

Rao K. Manibhushan : *Recent Developments in Biocontrol of Plant Pathogens*, Today & Tomorrow, Delhi, 1996.

S.K. Handa, N.P. Agnihotri: *Pesticide Residues : Significance, Management and Analysis*, Research Periodicals, Delhi, 1999.

Sathyanarayana B N : *Plant Tissue Culture : Practices and New Experimental Protocols*, I K International, Delhi, 2007.

Trivedi P.C. : *Plant Tissue Culture and Biotechnology*, Pointer Pub, Delhi, 2010.

Vanangamudi, K.: *Principles and Methods of Plant Breeding*, International Book, Delhi, 2005.

Verma, L.R. and Sharma, R.C.: *Diseases of Horticultural Crops: Fruits*, Indus, Delhi, 1999.

White, G.F.: *Natural Hazards: Local, National, Global*, Oxford University Press, New York, 1974.

Woolfe Jennifer A.: *The Potato in the Human Diet*, Cambridge, Cambridge University Press, 1989.

Yoshida T.: *Cultivation of Citrus Genetic Resources for Evaluation of Characteristics*, Tokyo, JICA, 1996.

Index

❑❑❑